REVISED 2ND EDITION

How to Be Your Own CONTRACTOR AND SAVE THOUSANDS on Your New House or Renovation

While Keeping Your Day Job

– with Companion CD-ROM –

Tanya R. Davis, Updated & Revised by
Billy Calvert and Martha Maeda

HOW TO BE YOUR OWN CONTRACTOR AND SAVE THOUSANDS ON YOUR NEW HOUSE OR RENOVATION WHILE KEEPING YOUR DAY JOB: WITH COMPANION CD-ROM REVISED 2ND EDITION

1405 SW 6th Avenue • Ocala, Florida 34471 • Phone 800-814-1132 • Fax 352-622-1875
Website: www.atlantic-pub.com • E-mail: sales@atlantic-pub.com
SAN Number: 268-1250

Library of Congress Cataloging-in-Publication Data

Maeda, Martha, 1953-

How to be your own contractor and save thousands on your save thousands on your new house or renovation while keeping your day job : with companion CD-ROM / By Martha Maeda. -- Revised 2nd Edition.

pages cm

Includes bibliographical references and index.

ISBN 978-1-60138-942-8 (alk. paper) -- ISBN 1-60138-942-6 (alk. paper) 1. House construction--Amateurs' manuals. 2. Contractors--Selection and appointment. I. Title.

TH4816.D3347 2015

692'.8--dc23

2014049821

Printed on Recycled Paper

Printed in the United States

A few years back we lost our beloved pet dog Bear, who was not only our best and dearest friend but also the "Vice President of Sunshine" here at Atlantic Publishing. He did not receive a salary but worked tirelessly 24 hours a day to please his parents.

Bear was a rescue dog who turned around and showered myself, my wife, Sherri, his grandparents Jean, Bob, and Nancy, and every person and animal he met (well, maybe not rabbits) with friendship and love. He made a lot of people smile every day.

We wanted you to know a portion of the profits of this book will be donated in Bear's memory to local animal shelters, parks, conservation organizations, and other individuals and nonprofit organizations in need of assistance.

– Douglas & Sherri Brown

PS: We have since adopted two more rescue dogs: first Scout, and the following year, Ginger. They were both mixed golden retrievers who needed a home.

Want to help animals and the world? Here are a dozen easy suggestions you and your family can implement today:

- *Adopt and rescue a pet from a local shelter.*
- *Support local and no-kill animal shelters.*
- *Plant a tree to honor someone you love.*
- *Be a developer — put up some birdhouses.*
- *Buy live, potted Christmas trees and replant them.*
- *Make sure you spend time with your animals each day.*
- *Save natural resources by recycling and buying recycled products.*
- *Drink tap water, or filter your own water at home.*
- *Whenever possible, limit your use of or do not use pesticides.*
- *If you eat seafood, make sustainable choices.*
- *Support your local farmers market.*
- *Get outside. Visit a park, volunteer, walk your dog, or ride your bike.*

Five years ago, Atlantic Publishing signed the Green Press Initiative. These guidelines promote environmentally friendly practices, such as using recycled stock and vegetable-based inks, avoiding waste, choosing energy-efficient resources, and promoting a no-pulping policy. We now use 100-percent recycled stock on all our books. The results: in one year, switching to post-consumer recycled stock saved 24 mature trees, 5,000 gallons of water, the equivalent of the total energy used for one home in a year, and the equivalent of the greenhouse gases from one car driven for a year.

Table of Contents

Introduction

Welcome to the world of self-contracting. Whether you are building your first home from scratch, remodeling a home you already own, or investing in property to fix up and flip, this guide will help you reach your goal. Building a home has become increasingly complex, but by developing a straightforward, organized way to handle each step of the process, you can easily create the structure you always dreamed of, from beginning to end. You will be able to construct a beautiful and functional house that will bring you joy for years to come. Best of all, you will have the satisfaction of having built it yourself.

You might be reading this book because you have a vision. Whether your dream is to remodel your current home or build a new home, you want "something different". Perhaps it is more space, more light, or a better view. You might have attended scores of open houses, only to find that none of them quite suit you.

After realizing that there is no perfect house on the market, many people think about tackling homebuilding themselves. The idea raises many questions. Can I do it? Is it too confusing and mysterious? Is it too difficult? And, most important, will it cost me more money than if I hired a contractor?

This book helps you answer those questions. Decide whether the homebuilding process is right for you. Turn your dream into a reality — a custom home that you can be proud of.

Every owner-builder has three objectives. The first is to build his or her dream home, or transform a current dwelling into the perfect house. The second goal is to save money, both by self-contracting and by selecting supplies carefully. The third is to save time. By streamlining the construction process, you can reduce time on the job and spare yourself endless frustration.

This book addresses all three of those goals. It shows you how to begin, and provides you with checklists for the major steps of the process. It helps you determine who carries which responsibilities, and how to make your site safer and more organized for your subcontractors (subs). It explains the contracting process and how to manage subs and materials — people and things — efficiently.

Whether your goal is to build one house or many, this book is a handy take-along reference, answering hundreds of questions to help you become a successful owner-builder.

Your Objectives

Learn how to lower your costs, save time, and eliminate delays and confusion. Instead of detailed step-by-step instructions for building a foundation, this book gives the six general steps for creating the foundation, lists possible mistakes, explains the most popular kinds of foundations, and describes how to site foundations on varying slopes.

Every chapter suggest ways to save. Chapter 10 is devoted to getting more for your money when you purchase materials. Chapter 7 explains the bidding process: how to find contractors to ask for bids, how to select and eliminate possible contractors, and how to get contractors to do the job for less.

This book contains all the information you need to manage your own residential construction project. Whether you intend to work on the home yourself, or never plan to lift a hammer, this book guides you through the contracting process. By following the steps outlined here, you can organize the project and complete the construction process quickly.

Learn which tasks are best left to professionals, and how to identify and avoid problems encountered by other homeowners. Study the best floor plans. Learn how to choose a site and position the house on it. Create maximum enjoyment for your family while streamlining the construction process, so that you can spend less time on it.

When you build your own home, you become intimately familiar with every detail of the house. You can arrange the floor plan to suit your family's lifestyle and activities. You can move the garage to the other side of the house, or even to the back. You can choose the thickness of the outer walls, and the kind of wood your beams are made from. You can use high-quality, durable products that will save you time and money in maintenance and repairs. The average homeowner who asks a builder to build a "custom" home has little to do with the details, choosing only square footage, paint colors, carpeting, and perhaps some of the kitchen features. By comparison, when you are finished with your home, you will know every component. You will know why you used specific materials, and you will feel much more satisfied with the house.

Many people believe that because they and their spouses work 40 hours a week, there is no way for them to oversee the construction process. This is not true. It is possible to manage the project using the phone and email, and spending weekends and evenings at the job site. You can save money on the construction of your own home, and keep your day job.

Your ultimate success will depend on your own management and do-it-yourself skills. Residential construction projects normally involve 20 to 30 specialty tradesmen; each has his own specs, terminology, and expectations. In this book you will learn the main concepts for each trade so that you can ask intelligent questions, gather bids, and select the right sub with confidence.

The Focus

This manual concentrates on two important aspects of managing your own construction project:

- How to save money
- How to juggle building or remodeling with your own job and family responsibilities

Each section of this book includes helpful tips from real people. I have built and remodeled several homes, and interviewed contractors across the US who shared their tips and suggestions. Each contractor gave me the name of at least one other contractor who was happy to answer my questions. I have wandered through construction sites, and spoken to many acquaintances who built and remodeled homes. I researched building extensively, looking for ways to save money.

The appendices at the end of the book list helpful websites, magazines, and other materials to help you learn how to be a builder. Talk to builders, subcontractors, and suppliers, and read as much as you can. The more information you have, the more confident you will feel about taking on the task. Relax, have fun, and enjoy building your new home.

CHAPTER 1

Are You Ready to Tackle the Custom Home Process?

You do not have to be wealthy to own your dream house. The secret is to build it yourself, or to remodel your present house into the custom home you always wanted.

Whether they are perpetrated by the construction industry or uninformed individuals, myths about contracting plague the do-it-yourself world. Many would-be builders are discouraged because they think that contracting is closed to private individuals. Building your own home is not only possible, it is a smart choice. By building a home, you save up to 35 percent of the cost of purchasing a previously built home. You are able to create the house you have always dreamed of. By building it yourself, you can create a structure that incorporates your priorities. Best of all, building your own house is fun.

Can You Design Your Own Home?

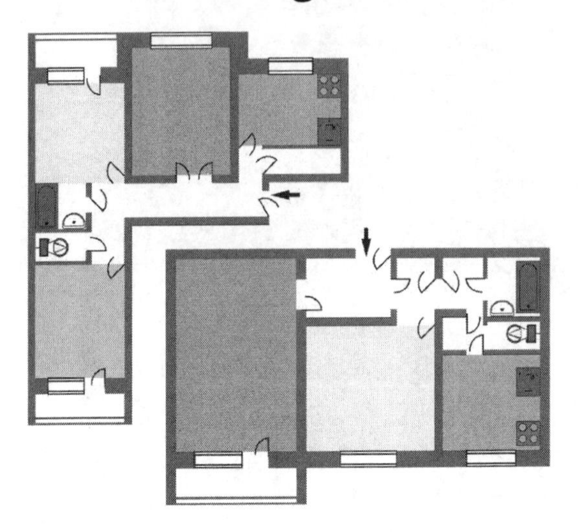

One of the main objections to building your own home is fear of creating a design. You have been in and out of homes all your life. By now, you have developed a clear sense of what you like in a home, and what you dislike. Who is better qualified than you to decide how your rooms will flow, what types of rooms you need, how much storage each one should have, and what view you see from the windows? Building your own home gives you the freedom to create exactly what you want.

If you suspect that you lack good taste or a sense of design, you can still have exactly the home you want, and build it yourself. Hire an architect to create the design you envision. By using an architect, you will avoid creating a home that is so unique that it cannot be sold later. An architect can also help you identify potential problems in your dream house design. If you cannot form a clear picture of what you want, the architect can begin a design and modify it to suit your taste. He or she can help you figure out what type of home is best for you.

Myths About Building Your Own Home

Myth No.1: High Materials Costs

One of the most common construction myths is that contractors get better prices on materials because they buy in bulk. This is not necessarily true. Most contractors do get special pricing from their suppliers, but you can also find good deals by shopping around. Contractors are also middlemen, and often mark up the materials instead of charging you the price they pay to suppliers. If a contractor sends you to his lumberyard, where you get a quote of $18,000, talk to other builders, and get estimates from their suppliers. You might save a few thousand dollars by being a savvy shopper.

Myth No. 2: You Have to Get a License

Another myth is that owner-builders are doing something illegal. You must own the land, and you must follow the local laws by getting all the right permits, having inspections, and eventually occupying the house, but nothing restricts you from building (except in certain restricted subdivisions). In most states, owner-builders do not need a specific license, and no special licenses are required unless you are attempting some of the more technical trades like electrical work, plumbing, or heating and air conditioning. Investigate the requirements for getting licensed as a builder in your state. One woman discovered that all she had to do was take a one-day class, which gave her the credentials required by lenders. Having a building license could help you get better prices and more funding.

Myth No. 3: Subcontractors Hate Owner-Builders

A sub who agrees to work for you clearly does not have a problem working for private individuals. If a sub does not show up for a job, it is not because you or your demands are unreasonable; it is because that subcontractor is unreliable. It is likely that he has failed to show up for other employers too. This book explains the hiring process, so that you can create a good relationship with the subcontractors you hire. It is important to interview the subcontractors and check their references before you hire them, so you can be sure that your workers will show up when they are supposed to.

Myth No. 4: You Cannot Do It

You might not know everything you need to know to build a house, but there is no reason you cannot contract your home. Contracting involves management skills more than construction. You can always hire outside help, and many people are willing to give you advice when you need it. Your job is mostly to organize and document the entire process. You or your spouse will have to spend time at the job site, and you must pay close attention to detail. There will be something to do every day of the project, and most nights you will go to bed thinking about what you need to do next.

If you are wondering whether you can manage your own project, the answer is: of course you can. Thirty-five percent of the homes in the US are custom homes, and they are not all built by professional builders. Let us look at some of the things that will make your project run more smoothly.

What It Takes to DIY

To build or remodel a home, especially if you are doing it while keeping your day job, you must be willing to learn. If you have never participated in the construction industry, or if you have knowledge of only one area, you must learn new terminology, the specifics of building, and local regulations. You will need to read everything you can find about building. Read books like this one and the ones listed in the bibliography, and read magazines about building, including Home, Home and Architectural Trends, Taunton's Fine Home Building, and This Old House.

Besides reading about building and design ideas, study the local codes and zoning regulations, and be sure that you comply with them. You also need at least some understanding of blueprints and their significance in the construction process.

Apart from reading and learning about the construction industry, you need only average organizational skills to act as your own contractor. Skills in the various building trades are a bonus, but not a requirement.

Your Responsibilities

Your main job as a contractor is planning. This involves scheduling workers to be on the site to perform certain tasks, and coordinating the supplies and equipment that they might need. If you have hired someone to operate a backhoe, for example, make sure you have rented and received the backhoe before he or she arrives to work. You do not want to pay people to stand around and wait for equipment. The better organized you are, the more money you will save.

To keep the task from overwhelming you, break it down into a series of steps, and proceed with each step as if it is a single project. For help with organizational skills, read Organizing for Dummies, by Eileen Roth and Elizabeth Miles. The National Association of Professional Organizers (NAPO) website, www.napo.net, helps people get organized.

The second responsibility of a contractor is to estimate costs. You will have to remain within your budget if you want your building loan to cover the cost of construction. Otherwise there will be cost overruns, which will force you to look for money in other places like your assets or savings account. Set aside additional funds for emergencies or any problems that occur. You do not want a cash-flow problem to keep you from finishing your house on time.

One important responsibility of a contractor is to find good subcontractors (subs). If you gather a roomful of do-it-yourself builders, you will hear plenty of complaints about subs like, "They do not show up." "They are always late." "I have to stand over them." Finding good subs is the key to getting your job done on time. If you are careful to hire the right people, they will do their jobs without close supervision, and you can spend less time on site. Chapter 6 explains how to hire good subs.

You will be expected to organize the construction job. As the contractor, you are the boss. No one will do his or her part without a directive from you. If you are well organized and stick to your budget, the job will go more smoothly. The workers will follow your lead, and they will complete their jobs on time without wasting resources.

Your Attitude

By nature, a contractor's job requires having certain expectations of the workers and suppliers he or she works with. Your attitude determines the behavior of the subs you hire. Your attitude also affects the way that you feel about the job — and about yourself. By communicating well with everyone involved, you can make sure that your expectations are understood. You will also circumvent many problems before they arise. No one wants to work for a rude boss. Projecting a positive image boosts morale, and compels your workers to be effective and efficient.

Can I Save Money?

You are sure to have heard at least one or two fallacies about homebuilding. For example, you may have heard that owner-builders do not save any money compared with the price of contractor-built homes. This is false. If you survey ten people who have built their own homes or have done their own extensive remodeling, you will find that every one of them saved money. Many will say that they saved around 50 percent.

You can save money by building your own home. This is understandable, because when a contractor builds for you, he is doing it to make money. His or her work always includes a "salary", though the structure of the contract does not call it that. Most builders mark up their jobs by at least 10 percent, so 10 percent of the cost is the absolute minimum that you will save.

Ten percent of a $300,000 home is $30,000 — quite a savings, especially when you consider the amount you would pay on that in interest over the life of your loan. If you use the tips in this book as a guide, you could save more. If you save 35 percent on that same $300,000 home, for example, that is $105,000 in savings because you decided to build the home yourself.

Now let us look at remodeling. Your total budget will probably be smaller when remodeling, but the savings can be astronomical. Suppose you own a home worth $298,000. You decide to remodel, converting your garage into another bedroom. While you are at it, you enlarge and opens up your kitchen with a 560-square-foot addition. Acting as your own contractor, you turn your three-bedroom home into a four-bedroom with a new kitchen addition. You spend about $66,000 on the project. You have saved between $6,000 and $7,000 by overseeing the project yourself.

You decide to refinance the home after the addition is complete, so you order an appraisal. The house that in your mind was worth $364,000 ($298,000 plus the $66,000 addition) appraises at $413,000. The value of your house has increased by $115,000, almost double the expense of the remodeling project.

If you find that you are not ready to build your dream home, there are advantages to remodeling an older home instead of building a new home. You can remodel the house you own, or look for a home in the neighborhood where you want to live that needs to be fixed up. This type of house is often advertised with a description, like a "diamond in the rough", that indicates a less-than-perfect appearance. Sometimes a low-interest loan that can be assumed exists on the property, making an older home especially attractive. If you are in a hurry to move, a remodeling project can be completed months before a complete new build would be. Some areas offer tax incentives for remodeling.

Common Problems

Certain problems crop up again and again with custom home building. Some of these are due to "other people"— people who tell you that you cannot do what you are doing. Some problems arise because you are trying to juggle a home life, a job, and a large construction project. Some are simply due to lack of experience. These are some of the most common problems:

- **Lack of a license** — Do you need a license? Normally, special licensing is not required if you are building a home for yourself, but check with your local building inspection department to be sure. Licenses are required for specialty trades.

- **No flexibility** — If you are building a home from the ground up, you need to spend at least part of every day on the job site. This could mean that you or your spouse will have to set aside the time to be there. If your job and family life do not allow you to spend time at the job site, find alternate ways to handle problems and supervise your employees. For example, if a sub phones with a question at 1 p.m., make sure someone can get there within the hour to look at the problem.

- **No time** — If you find that the building project is becoming too stressful and you have no time for your family, consider paying a builder for assistance. The builder can help you by attending the site while you are at work, lending you his subs, and even helping you to get a construction loan. Builders will often do this if they are not busy, because being present on your project might secure them other buyers.

- **Treating the project as if money is no object** — Everyone has financial limitations. To stay within those limits, set the budget before breaking ground, and then determine to stick to it. The best approach is to visit lenders first, before you have many details in your dream house design. When you know your financial constraints, you can make faster, more reasonable decisions.

The Advantages

There are many advantages to building your own house. Most people who have done it say that they enjoyed having total control over the quality of construction and the materials. Many contractors have a certain method they like to use, or a certain way of doing things; they do not necessarily change their system to accommodate the customer. By building yourself, you can "have it your way."

Most owner-builders save an average of 15 to 30 percent. This creates an instant home equity on the property, and whittles away some of those monthly mortgage costs. Because of the do-it-yourself savings, your payment will be lower than it would have been if you had purchased a ready-built home.

You choose the floor plans. Often builders have a collection of standard plans, which might meet your needs. If the builder does not have standard plans, or if you want to customize a floor plan, he or she will accommodate those changes -

at a cost. When you build your own home, you can do the research described in Chapter 5 and find a plan that is exactly right for you.

Another advantage to building your own home is that you select the building materials that are used. As you research the construction process, you will learn about types of foundation, subflooring, and plumbing. For example, you will discover that manufacturers recommend "optimal" sizes of furnaces, but that there are "better" sizes for specific circumstances. By selecting your own materials, you can control the quality of the house, and emphasize the elements that are most important to you.

The best part of building your own home is the satisfaction you feel when it is complete. Forming an idea of what you want your home to look like, and then working hard to make that idea a reality, will be one of your biggest accomplishments. The completion of the project — the culmination of all your hard work — will be the proudest moment of your life.

Create a Workbook

To get started on your project, use a workbook. This can be a three-ring binder, with or without plastic pocket sleeves. Pocket sleeves allow more flexibility if you decide to move things around. Alternatively, use an expanding folder with 13 sections. Set up a file drawer for the project in your home office; the paperwork generated will probably occupy two entire drawers.

In the accordion file or notebook, label these categories:

- Dream project
- Design plans
- Contracts
- Communication
- Financial Information
- Invoices
- Lot
- Materials

- Receipts
- Permits
- Subs
- Warranties

Under the "dream project" section, start saving magazine articles, paint-chip colors, and other tidbits you have gathered as you have thought about building your own home. It will be helpful later to have these ideas on hand. Having all these images on hand allows you to mix up designs, and see how your colors and shapes work together. Keeping everything in one place allows you to make informed decisions.

True Financial Ability

After setting up a filing system, the first thing you must do is determine the budget for the project. Nothing can be done until you are certain how much money you can spend on the house. This will largely be decided by the bank, but it would be make an appointment with your accountant to assess your finances first.

Many people simply visit lenders and they choose the one who offers the "best" loan package. The amount of the loan you are offered often results in a higher monthly payment than you are comfortable with. To plan for a loan amount that fits your financial circumstances, define the factors that will have an impact. These include:

- How much cash you have on hand
- How capital gains may affect you
- Other assets
- Anticipated expenses, like upcoming college tuition
- Your long-term investment plan
- How long you plan to own the home
- Expected appreciation on the property
- The tax implications of your loan package (interest and points)

After you have considered these factors, talk a loan officer. Chapter 4 explains how to apply for loans, and what type of loans are appropriate for owner-builders. A loan officer can help you determine the loan amount that you can afford. You can use some of the online calculators at www.mortgage-calc.com to find out the amount of money you might qualify for.

Set a Budget

Once you have determined the loan amount, set a budget for the construction process. Your budget should not be 100 percent of the loan; no matter how careful you are, something always causes building projects to go over budget. Set the budget at 90 percent, or better yet, 80 percent of your total budget. Setting aside emergency funds will prevent hang-ups that could destroy your project. Running out of money could force you to sell assets, or even to delay the project indefinitely.

Decisions, Decisions

Your first big decision is where you are going to build your house. If you are looking for land, research and explore all the options. You cannot find just the right land sitting at home in front of the computer. You have to go out and hunt for it. Expect bad driving directions, mislabeled streets, and properties that are not at all what you imagined. Consider it an adventure — sit back and enjoy the ride. You will know when you find just the right spot.

Choosing a Lot: What to Look For

There is a difference between buying a lot and buying land. A lot, or finished lot, is already prepared for a home to be built on it. It has some or all the utilities, and the ones that are not present on it are accessible.

Land, on the other hand, is not ready for building. Raw land might be harder to finance — or the financing could be more expensive — because it requires extra work. If you have set your heart on buying raw land, consider looking for private financing, or ask the seller to carry the loan. Allow more time and money for the building process; raw land will require special permits, roads, utilities, and possibly drilling a well.

Do you select the lot first, or the style of home first? For now, evaluate the lot as if you have already determined that the house you are going to build will fit on it.

Money-saving Tip:

If you want to save money, purchase a "finished lot" instead of raw land. Lots can be purchased with less money down, they are easier to finance, and you will spend less time prepping the property for construction.

Lay of the Land

When looking at lots and land, there are so many factors that the choices may be dizzying. Many of the choices are extremely personal, and will be determined by your lifestyle. Your "perfect" lot might be one your best friend would never consider.

What does it mean when a lot is labeled "buildable"? Even if the city, county or state in which the land is located has deemed it buildable, will it be economically realistic to build there? Often when a site is referred to as buildable, even when the seller uses that word in the advertising copy, there may be good reasons not to build on it. To determine whether a site is buildable for you, make sure that it meets every requirement at every level of government. Dig for information. One experienced home builder told me that he had learned to go to the government offices and ask repeated, non-stop questions — then go back the next day, and ask some more. He had learned the hard way that if you do not ask exactly the right question, you might not get the answer you need.

It is not up to the government office workers to solve your problems. You are responsible for your own purchase, design, and solutions.

When evaluating a lot, walk every inch of it. Study the lay of the land, note where drain pipes have been placed, and look for soft spots. If there are trees, many of them might have to be removed for construction. If there are rocks, you might have to excavate or blast them. Also consider the following:

Slope

If your family spends a lot of time in the yard, avoid a lot with a steep slope. If the lot you are considering is sloped, study the way it drains — does the water drain toward the house site? Will it be difficult to mow? Will you have a hard time driving up the driveway in the winter? Does water collect on the property? Also consider the implications of a lot that is too flat. If the slope is less than 2 percent, it may have drainage problems

Look for land that is fairly level or that slopes gradually upward from the road. If you particularly like a location, but the slope is not exactly what you want, it is possible to grade it. Add the cost of grading to the total purchase price of the lot. In addition, if you remove a lot of dirt, you will have to pay to have it hauled away.

Privacy

Most people enjoy some level of privacy in their yard. If you select a home site on a busy street, you will want privacy in the rear, and possibly on the sides, depending on your personal lifestyle. Corner lots might feel less private, and they also have a shape that will complicate the placement of the home. Also, make sure the lot is big enough to leave a comfortable amount of yard space between your house and your neighbor's lot. Trees or fences can help create a sense of privacy, but you will need enough room for them in the yard.

City versus Country

Are you more comfortable in the city or the country? The old saying about the three most important factors being location, location, and location is not far from the truth. You are going to live there. Land outside the city will have fewer local taxes; land inside the city might have better access to roadways and highways. Country land will be quiet and peaceful, but city land will be close to nightlife, restaurants, and shopping centers. Think about your lifestyle and what location will most improve your quality of life.

Another consideration when determining the location of your house is how much time you are willing to spend on maintenance. Mowing and landscaping a couple of acres costs money and time. For some homeowners, the extra time spent is worth it. Others would not dream of spending their weekends doing yard maintenance.

Subdivisions

One of the advantages to choosing a lot in a subdivision is that resale is usually fast and profitable. In addition, home values will be somewhat stable because the homes are governed by covenants. Before purchasing in a subdivision, get a copy of the covenants. Be sure that they allow owner-builders, and that they protect against the building of much lower-priced homes than the one you are planning. Do the covenants dictate when you can or must build? Do you have control over the timeframe, or do you have to build within a certain time after purchase?

Also, pay close attention to homeowners' policies and subdivision rules. Many subdivisions have rules governing the appearance of your property and any additional work that you wish to do to the property. Subdivisions offer a more social community atmosphere, and if you hate yard work or some of their other required maintenance, you can find a neighborhood teen to do it for you.

Style of Home

You probably have a favorite style of home, or a type of home that you always dreamed of owning. Your dream house file might have pictures of many similar homes — so many that you realize this is your "type" of house. When looking at land, consider this type of home — the style you plan to build — and compare it with the homes that are in view of your proposed site. If you are building a log home, you do not want to purchase a lot in a subdivision full of Italianate structures. Likewise, if you are building an Italianate structure, you do not want to put it in view of a subdivision full of manufactured homes. While there is nothing wrong with having a house that stands out, you do not want to feel like your home is out of place.

Consider Resale

Many people build a home with the intention of staying in it forever. Others know that they will get bored, be transferred, or that the family size will change, and the dream home will have to be sold. Consider the resale possibilities when purchasing your land. The most difficult land to resell will be:

- Located in a flood plain
- In or near a declining neighborhood
- Near airports, railroad tracks, landfills, electric lines, power easements, radio towers, and major roadways

These factors not only affect the resale value of the home, they also influence your life in that home. Late night sounds from trains and airplanes may drive you crazy, and who wants to deal with a flooded basement? You do not want to own the most expensive home in a low or mid-range neighborhood. It can make your home more difficult for a future buyer to finance.

Easements, Encroachments, and Other Criteria

The different government authorities have criteria that have to be met, both to lessen the impact on the environment and to satisfy the residents of the community. When you find a piece of land that you like, check with the land assessor associated with the county to determine its legal status. He or she can show you the legal description of the lot, and tell you whether there are any easements that could affect your use of the land.

Zoning

There are several designations of zoning, each of which has a different meaning and certain limitations. The zoning tells you:

- The minimum lot size you can use
- What kind of structure you can build
- How many units can exist on the lot
- What the property can be used for

Zoning information is available at the local government office. Go to city hall if the property is in the city, or to the county offices if the property is in the county. Most lots for building houses are zoned R-1 (residential with one unit). A higher number means that more units can be built on the property. If you are looking at agricultural land, you might see a designation like RA-3. Rather than referring to the number of units, this zoning means that the residential agricultural property must be at least three acres in size. So, if you purchased five acres that was zoned RA-3, you could not subdivide later and sell off part of the property for building; each lot has to be at least three acres.

Zoning can change seemingly on a whim. The local government is not required to check with you, obtain your approval, or even notify you of zoning changes ahead of time. The best way to check on zoning is to visit with the local government office to see if there are zoning meetings scheduled for your area. While you are there, find out the zoning of the surrounding properties and whether there have been recent changes to them. Also, ask if there are any special tax assessments you should know about. Special tax assessments are charges that the government can place against your real estate for public projects. These include water or sewer lines, paving, parking structures, streetlights, and other assessments that property owners are required to fund.

A lot in a subdivision might be subject to more building restrictions, in the covenants, conditions, and restrictions, than a freestanding lot. Obtain these restrictions through the homeowners' association or through a title company. Examine them for setback limits, style limitations, height limits, and design review guidelines before putting an offer for the lot in writing.

Buildability

Determine whether the site is truly buildable — whether the soil will percolate, whether the land might flood, whether a septic system will be allowed. A "perc", or percolation test, is a soil test to see how fast water will drain through the soil. Perc tests are performed by septic specialists to see whether a septic system will function properly. The county then approves the septic design.

If there are wetlands, creeks, or lakes on or near the property, certain environmental studies are required. If the site involves a hillside, a geotechnical analysis determines whether it is a seismic hazard or landslide area.

Checklist for Buildability Verification

☐	Legal description
☐	Zoning designation
☐	Lot size versus setback
☐	Encumbrances, encroachments, right-of-ways
☐	Availability of sewer or an approved septic design
☐	Availability of electricity (also find out the requirements)
☐	Availability of natural gas (also find out the hook-up requirements)
☐	Availability of police and fire protection
☐	Driveway accessibility - does it require an easement?
☐	Storm drainage requirements for the local area
☐	TV and cable service requirements
☐	Geotechnical analysis - is one needed
☐	Seismic area review - is it needed
☐	Covenants, conditions, and restrictions if any

ITEMS THAT AFFECT COST	
☐	Extensive water, sewer, or electrical hookup
☐	Fire marshal's requirements (i.e., fire truck turnaround)
☐	Access roads
☐	Storm drains or other runoff control
☐	Wetlands or other environmental impact
☐	Removal of hazardous wastes, if present

Budgeting for Land Purchase

How much should you spend on land? It is difficult to determine early in the game how much you should spend on the land. A good guideline is 25 percent or less of your total budget.

Money-saving Tip: How to Save on the Land Purchase

When purchasing land, it is of paramount importance to know what the property is worth. This is part of your due diligence process – the research you complete before the offer. As soon as you can, find out what the property's market value is by looking at the prices of the most recently sold land in the same area. If the lot is located in a subdivision, look at the last properties sold in that subdivision. How were they like yours? How were they different? What was the price? By comparing the desirability of your lot with those that have recently been sold, you can get a good idea what the market value should be.

Make the property work for you financially by purchasing it first and paying it off. You can then use the land as collateral to borrow money to build the house. This is called leveraging the property, and is a good way to ensure that you will get a construction loan.

Checklist for the Lot You Are Considering

☐	Does it allow for a basement either by sloping or good drainage?
☐	Is it on the sewer line already? If so, is the lowest drainpipe placement going to be higher than the sewer line?
☐	Is it in a desirable location with a safe atmosphere?
☐	Is it in a low-traffic area?
☐	Will the shape be suitable once a house is built on it?
☐	Is it in a flood plain or full of subterranean rocks?
☐	Is the base soil or shale? Is it solid?
☐	How attractive are the surrounding homes?
☐	What is the view like?
☐	Is the entire lot usable?
☐	How is the area zoned? Is that suitable for you?
☐	Are underground utilities available (phone, electric, cable, and television)?
☐	Is the lot private?
☐	Will most of the trees and shrubs remain once construction is under way?
☐	Is it accessible to major highways?
☐	Is it convenient to shopping, schools, parks, etc.?
☐	How much are taxes per year? Is it inside or outside the city limits?
☐	Is it located in an area of growth or decline?

Home Design Checklist

☐ Style: ____formal ____casual

☐ Number of stories_____

☐ Number of bedrooms____

Every bedroom should have two means of escape in case of fire, a closet and easy access to a bathroom, and room on one wall for a king- or queen-size bed. You should not have to walk through bedrooms to get to other rooms.

☐ Great room ___yes___no

☐ Living Room ___yes___no

☐ Media Room ___yes___no

☐ Mud Room ___yes___no

☐ Number and location of bathrooms _________

☐ Location of the laundry room _______

☐ Location of the master bedroom ________

☐ Regular height ceilings___ or tall ceilings ___

☐ Eat-in kitchen____

A kitchen needs a clear walkway of 32 in. at each door, and 42 in. for the work aisles. At least 60 in. of wall space is required for cabinets, with the cabinets being 12 in. deep and at least 30 in. high. Sinks and food prep centers must not be separated by tall cabinets or the refrigerator. The entire "work triangle" (see Chapter 12) should be 26 ft. or less. Each leg of the triangle should be between 4 and 9 ft.

☐ Formal Dining Room ____

Allow at least 10 by 12 ft. for an 8-person dining room.

☐ Location of the storage areas _________________

☐ Number of cars in the garage _____

☐ Location of hallways _________________

Hallways should be a minimum of 36 in. wide, and be as short as possible to save money.

Chapter 1 Checklist

☐	Create a workbook with all the proper categories.
☐	Examine your financial ability, including current assets, debts, future obligations and income.
☐	Set a budget for the land and house.
☐	Select a lot, considering: • Lay of the land, including slope and drainage • City versus country • Subdivision versus acreage • Style of home you desire • Resale potential • Zoning and restrictions
☐	Verify buildability.
☐	Complete lot checklist.
☐	Complete design consideration checklist.

CHAPTER 2

Plans

One reason for building your own home is that you can design it yourself. You might have a clear idea of the style and design of the home you are going to build. If you do not, there are several ways to create house plans. Even if you are going to hire an architect to design your house for you, study house plans and familiarize yourself with floor plans and how they translate into "real" rooms. Determine the exact floor plan that will work best for you and your family.

By now, your "dream house" file contains a list of the requirements for your house. If it does not, make the list before investing time looking at house plans. Your list should include what types of rooms you want, their approximate sizes, and the number of rooms. When you find a floor plan that you think you like, check it against your "must-have" list.

Where to Find House Plans

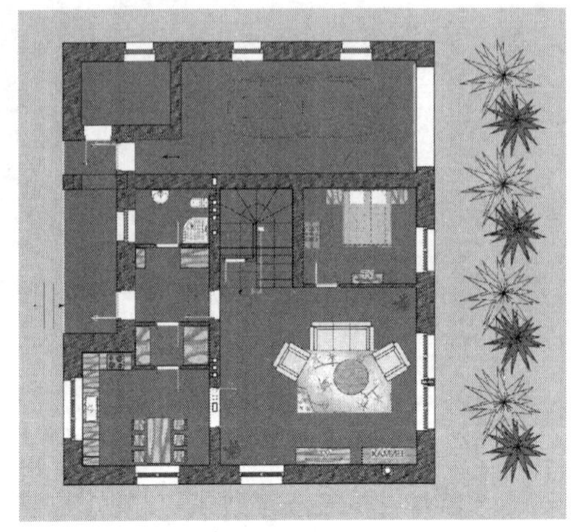

You can find house plans in most libraries and bookstores, on the Internet, and in building, renovation, and home interior magazines. Books and CD-ROMs of house plans are sold online or at the local builders supply store.

House plans are sold individually or in sets. When you have decided on your house plan, you need a copy for yourself, one for your lender, and one for your contractor or subs. If you choose to put the project up for bid, each bidder will want a set of the plans to look at. Most home builders order at least six sets of plans.

Money-saving Tip: Copy Your Plans

You can get copies made of the set of plans for about $10 each. If ordering extra sets of plans is expensive, save money by having them copied at a local office supply store or copy center, unless they are copyrighted.

The advantage of ordering one of the thousands of pre-drawn house plans is that, if you make no changes to it, you do not need to hire an architect or designer for your project. If you do have to make a change, hire an experienced draftsman, designer, or architect. Requesting changes to a pre-drawn plan will cost much less than having someone design the entire house.

Draw Your Own

Purchasing stock plans is the least expensive way to design your home, but you can draw your own plans if you prefer. To begin drawing your plans, purchase simple graph paper. Use a scale of ¼ in. to 1 linear ft. Draw the plan the best way you can, and then show it to your builder, architect, or designer. He or she can take your estimates and point out where you need to add or subtract items to reduce waste and save costs. Professionals can also point out parts of your design that may be impractical or inefficient. For example, staircases take up a lot of space, and a professional will tell you how to alter the design to maximize your floor plan.

Another way to create your own plans is to purchase a design kit. This kit will contain:

- Scaled grid sheets
- Appropriately sized furniture
- Parts, like windows, doors, and walls
- Landscaping items like trees

Design kit can be two-dimensional or three-dimensional, and may include items like plumbing fixtures, TVs, and appliances. A kit can make it easier for a professional to create the plans, so do not hesitate to use one. A kit makes it much easier to visualize the project as it will be when it is completed. A multi-dimensional view of your project will give you a more accurate idea of what the home will look like, and you might come up with new, better ideas.

Design Software

Computer Assisted Design (CAD) engineering and graphics design software guides you through the process of designing your home on a computer. You can rotate the drawings on the screen to see all the angles of the room, and create scaled drawings that show the proportions of the elements in your home. Drawings can be saved and printed. The designs are surprisingly accurate, and they are so professional that you can forgo the cost of having a professional complete the set of drawings. If you use CAD, take time to learn how to use the program before you start. If you do not know how to enter the correct dimensions, you may ruin the flow of your design.

Architect Versus Designer

Most municipalities require that plans meet certain building codes and standards. For example, in New York you must have the design certified if it is over 1,500 square feet; additions and remodeling over $10,000 fall under this requirement, too. This certification done by a design professional — either an architect or a designer.

Architects have more formal training than designers. They study at least six years in school, do internships, and then have to obtain architectural licenses. Architects also work with the building codes day in and day out, so they can save you time and money.

An architect is familiar with the special building considerations for your particular area. Ask to see his or her portfolio, and be sure that you use an American Institute of Architects (AIA) certified architect. Architect fees are structured as an hourly rate, a lump sum, or a percentage of the construction cost. Expect an architect to charge from 5 to 15 percent of the construction cost.

Designers can draw plans and also help you determine what materials are needed. Some home owner-builders like to draw the plans themselves, and then hire a designer. Another option is to purchase stock home plans, then employ a designer. You can save money by hiring a home designer who is not a licensed architect. A designer can also help you choose the most durable materials for the lowest cost. Search for a designer on the Internet, or look at the ads in home design magazines or your local newspaper.

Architects and Remodeling

You and your subs can handle remodeling projects that only address aesthetic changes, such as appliances, flooring, and countertops. However, anything that changes the footprint of house, like changing the shape of a room or adding rooms, requires an architect.

Some states prohibit unlicensed individuals from designing entire houses, so check your state guidelines before you begin work. Unlicensed persons might be allowed to complete a smaller job, such as a remodel or small addition to your home.

Building Green

Green building is the practice of reducing the impact of building on the environment. Key aspects of green building include using green building materials, from local sources whenever possible; generating at least some on-site energy; optimizing systems like heating and cooling; and recycling. Low-impact building materials include paints and adhesives that do not emit Volatile Organic Compounds (VOCs). VOCs are the cause of about 3,000 diagnosed cases of cancer each year. Using local materials eliminates emissions created during transportation, and also supports your local community. Energy-efficient heating and cooling will save money over time because your energy bills will be much lower.

Many green builders salvage materials from old houses — doors, windows, mantels, and hardware. This prevents old materials from ending up in a landfill. Using products made with bamboo or cork oak ensures that the wood you use is rapidly replenished. Bamboo grows quickly and is a beautiful, strong wood.

Close-up of bamboo flooring

Energy loads are controlled by using daylight for lighting, passive solar heat and thermal mass storage for heating, and by recycling hot water. These design features will save on your energy bill, and reduce your environmental impact.

Green building utilizes many interesting materials. One of these is straw bales, used for walls. They are covered with stucco and have an R-value of R-30 — nearly three times that of traditional wood stud wall systems. Higher R-values mean that the wall retains more thermal energy and is better insulated.

Green building practices include:

- Using recycled content, materials from sustainably managed sources, or materials that are natural, plentiful, or easily renewable

- Using materials that are found locally or regionally.
- Using items that have been salvaged from disposal and renovated, repaired, restored, or improved through remanufacturing
- Using recyclable or reusable materials, enclosed in recycled content or recyclable packages
- Using materials that are longer lasting than conventional products, so that they will need to be replaced less often
- Improving indoor air quality with low or nontoxic materials which emit fewer carcinogens or irritants and minimal VOCs. Air quality is also improved by using products and systems that are moisture resistant to prevent mold and biological contaminant growth.
- Using materials and components that require simple, nontoxic, low-VOC cleaning methods, and products that enhance the air quality or identify indoor air pollutants
- Maximizing energy efficiency by using materials, components, and systems to reduce energy consumption. This includes:
  - Passive solar design or other building shape and orientation that utilizes natural lighting
  - High-efficiency lighting systems
  - Properly sized and energy-efficient heating and cooling systems
  - Lighting equipment and appliances that minimize the electricity load
  - Alternative energy sources where appropriate
- Minimizing wastewater with low-flush toilets and low-flow showerheads and faucets. Often, green builders use recirculating systems for centralized hot water distribution, dual plumbing designs to recycle gray

water for toilet flushing, and rainwater recovery systems for landscaping irrigation. Green builders also use micro-irrigation to supply water for landscaping, and self-closing nozzles on hoses.

One objection to building green is the extra expense. Green building products are becoming increasingly affordable. Keep in mind that a cost now, such as solar panels, is recovered in a savings over time.

Green Building Room by Room

Bathrooms

The US uses 300 billion gallons of water every day, enough to fill Olympic-sized swimming pools placed end-to-end all the way around the earth. Water conservation may be a high priority in your area. If it is, consider installing low-flow toilets, showers, and faucets. Install a hot water recirculation pump with a timer. These recirculation pumps are convenient because you do not have to wait for the water to get hot; it is already in the pipes. Also, because the water gets heated based on a timer, energy is not wasted by reheating water that is not needed.

Kitchen

Approximately one fourth of the nation's largest water treatment facilities are in serious violation of pollution standards at any time. More than seven million Americans become ill each year due to contaminated tap water. If this is a concern to you, consider installing a distiller. These actually cost less than going to the store and buying bottled water; they use about 3 kWh of electricity per gallon. Distillers remove contaminants from the water you drink.

Electricity

Electricity uses up a lot of natural resources. Most of it is derived from coal and other sources that have a large carbon footprint. To reduce your carbon footprint, use "green" power, or solar energy.

Local power utility companies provide green power by buying solar or wind power and providing it to you. The cost is slightly higher, but some areas offer incentives or tax breaks to offset the extra costs.

Incorporating passive solar design into a new construction is easy. Passive solar design uses the sun to heat your home. It can work on its own, or be used in tandem with more traditional sources of heat. If you want to use passive solar energy, plan to face windows to the south to collect heat.

Solar systems can be on-grid (connected to the power utility) or off-grid. If you are building in an extremely remote area, consider off-grid systems with battery storage.

Indoor Air

Indoor air pollution is one of the top five environmental risks. It causes 14 times more deaths than outdoor air pollution. To reduce indoor air pollution in your new home, use low-VOC paint, seal duct work tightly, and insulate effectively. Visit www.epa.gov for more information about clean indoor air.

Green Floors

Earth-friendly flooring is referred to as green, natural, or nontoxic. Many consumers now demand sustainable flooring because it improves the environment and reduces toxicity levels in their homes. Earth-friendly flooring includes wood, bamboo, cork, natural carpeting, concrete, and natural linoleum.

- **Concrete** — Concrete is becoming more popular as a flooring choice. It is inexpensive and can be painted, scored, or stained. It can be cold to the touch, though, and presents a hard surface to stand on if you are on your feet a lot at home.

- **Carpeting** — Green carpeting refers to carpeting made of natural materials: coconut husks, sisal, sea grass, wool, and jute. Green carpets are low in VOCs, stain-resistant, and made from recycled materials. Be sure the backing on the carpet is also environmentally friendly.

- **Linoleum** — Linoleum is nontoxic and biodegradable. It is made from natural materials like linseed, rosin, or jute. Linoleum can last up to 40 years and is a great alternative to vinyl, which may contain harmful pollutants.

- **Wood** — Not all wood is earth-friendly. If wood is not forested properly, it may harm the environment. To purchase hardwood flooring that is green, look for products that have been certified by the Forest Stewardship Council (FSC). FSC-certified wood is available in all the popular species and some of the more exotic woods like teak. The FSC works to make sure that environmental pollution, habitat destruction, and wildlife displacement is minimal. If you choose hardwood for your floors, be sure to use low-VOC stains and finishes to reduce toxicity and the release of harmful chemicals.

- **Bamboo** — Bamboo is an extremely renewable resource and a popular choice for flooring. It grows fast and easily with little fertilizer, pesticides, or care. Bamboo makes a beautiful, durable floor that is water resistant. When you shop for bamboo, be sure your supplier knows its source; formaldehyde may have been used as a binder—he will know for sure.

- **Cork** — Natural cork is made from the bark of the cork tree. It is resistant to mold and microorganisms and it feels great under your feet. Plus, cork is reduces sound. Cork costs about the same as hardwood flooring and can be used for floors, walls, and underlayment.

Outdoors

To continue the green building concept on the outside of the home, use steel framing instead of wood; steel products today contain 60 percent recycled content, compared to the 40 old trees that would be required to frame a 2,000-square-foot home. Fencing and concrete products made from recycled materials are readily available. Urban compost and mulch made in your hometown, or nearby, can be delivered when you are ready for landscaping. Salvaged brick and stone make use of existing resources instead of straining the supply of virgin resources. Also, consider recycling your construction waste. Forty percent of the waste that goes into landfills is from construction; more than half of this is recyclable.

Green Remodeling

If you are interested in cutting the heating bill and reducing your carbon footprint, consider making some changes during your remodeling project. Approximately two-thirds of your energy bill goes into heat; half of that is wasted. To offset this waste, insulate the basement, attic, and garage. Insulate the water heater, if it is older, and wrap your exposed hot water pipes with insulation. Consider replacing the water heater and other appliances with ENERGY STAR® equipment to reduce costs. Make your house extremely green by using under floor heating and solar water heating, and supplementing your regular heating system with a wood stove or a pellet stove.

The US Green Building Council website (www.usgbc.org) and www.doityourself.com offer topic-specific green building articles. Learn about ENERGY STAR® ratings at www.energystar.gov. You can find green building supplies at www.greenbuildingsupply.com.

Placing Your House on the Lot

Siting your home is one of the most important aspects of building. The way your house sits on the land determines how sunlight enters your rooms, what sort of view you have when you look out the windows, and whether you have privacy or feel as if you are in a fishbowl. It is important to site the house in such a way that you will feel comfortable inside it.

Prior to deciding how to locate the house on your lot or land, get a site plan from the surveyor. This plan shows the dimensions, easements, and setback requirements, so that you can place the house properly. If you do not, you might encroach on someone else's property or on a public easement.

Study the property's drainage pattern. Look for potential problems, like rocks or springs that might be located under the soil. These must be avoided, which may mean that you have to set the house differently than you originally planned. Look at the way the property will drain after grading for a building. If you change the drainage pattern to flow onto someone else's property, you can be held liable for damages, so be sure that after you have located and graded the property, it does not drain onto the neighbors'. Make sure that the water on your property will drain into an unneeded space. For example, gardens, driveways, and patios could be damaged by too much water.

To lay out the corner stakes, first determine which direction the house will face. Most houses look best when seated parallel to the street, but it is perfectly acceptable to orient a house some other way — especially if you are concerned about sunlight for solar heating. Be sure that you honor the setback limits in the front, side, and back of the house. Measure the front line first, and then locate the back corners. Square the stakes by measuring the diagonals of the square—they should be the same. Then go back and measure all the setbacks again to be sure you are in compliance.

Contents of House Plans

House "plans" actually consist of several parts besides the floor plan:

A Plat

You, the designer, or a surveyor will create the plat. It is a map of your lot with the house drawn in. Its purpose is twofold; first, if your house will not fit on the scaled drawing of the lot — then it will not fit on the real lot. Second, the plot plan is required to get the building permit. To create the plat or plot plan, first draw the lot, then sketch in all the zoning setbacks and restrictions. Next, draw in the house.

Spec Sheet

The spec sheet is a list of the specifications for the materials to be used. It includes the type and grade of wood, block, roofing, concrete, siding, brick, insulation, and so on. There is a sample spec sheet in Appendix A. The easiest way to create the spec sheet is to use an existing one and add to it as needed.

Floor Plan

The floor plan shows the entire outside dimension, location of windows and doors, and plumbing fixtures. Electrical, heating, and cooling fixtures are not included in the floor plans, but will be drawn out on separate sheets. There should be one plan for each level of the house.

Foundation Plan

The foundation plan simply shows the house and the location of load-bearing components like the footers, basement slab, piers, and reinforcing rods.

Detail Sheet

This diagram gives the specifications for footings, framing, and a layout of the kitchen cabinets.

Elevations

The elevation sheets show the outside of the house from all sides. There are a front, rear, and two side views. An elevation sheet shows the windows and doors — location, as well as size — and indicates what kind of exterior materials will be used.

Special Plans

The set of plans may include a mechanical, electrical, and plumbing layout, or these may be done by the subs. Additionally, there may be other detailed layouts like trim, tile, windows and doors, and the roof.

Design Considerations

The lender might require you to build the house exactly as the plan was presented. Changes that seem insignificant can have a ripple effect. For example, adding a couple of feet in an upstairs bedroom will cause changes on the floors below — which can end up being extremely expensive. The best way to save money is to keep from making any changes once the plans are in place. Consider design changes now in the early stages. Take your time and examine all the possibilities. There will always be items that you realize you want to change, but these changes should only be superficial. Changing the structure of the house after construction has begun is unrealistic. Then, you will only be able to fine-tune details such as kitchen cabinets, or which way the doors swing.

Traffic Patterns

An acceptable traffic pattern pushes movement to one side, rather than circulating people directly through the center of a room. If your floor plan does this, consider whether simply moving a door toward a corner will solve the problem.

Storage

Some home designs — even designs regularly used by popular builders — do not have enough storage space. Add bigger closets in the bedrooms than you think you will need, and consider bumping out the garage by a couple of extra feet. Plan for storage in the attic or basement, and build a pantry into the kitchen.

Placement of Windows and Doors

Windows can let in unwanted heat, invite burglars, and turn your house into a fishbowl for the neighbors. Exterior doors give the first impression of the home; interior doors may block passageways or prevent large furniture from going in or out. Consider carefully every placement and measure the doorways. Interior doors should wing into rooms rather than the hall.

Convenience

Think about your daily activities. How well does the new floor plan accommodate the things you do every day? Pretend that you are walking and living throughout the new plan. How does it feel? How convenient will it be to do laundry or carry in the groceries?

Building for the Future

Families go through stages, and as a result, each family outgrows their home approximately every five years. If you can plan your house accordingly, you might save yourself from moving. If you have not yet started a family, consider building a three or four bedroom home. If you have several teens who will be off to college in a few years, build a style of home that allows you to close off a portion.

Potential Problems

Just as beauty is in the eye of the beholder, whether a building issue is a problem depends on you. You are building your house; if you put enough thought into it, you can make it exactly what you want. However, there can be hidden problems in a house plan. Here are some of the most common problems.

- **Deciding to build a certain house that you have seen in a magazine before finding a lot.** Finding a lot to fit house plans that you have already selected can end up being very expensive. Instead, consider finding a couple of potential lots first, then finding plans that will fit them.

- **Choosing stock plans that do not match your climate.** Stock plans purchased from a magazine or website do not take into account your climate or the nature of your lot. For this reason, you might want to hire a professional to make modifications or at least look it over.

- **Wasted space.** Long hallways are the biggest cause of wasted space. Once you have the basic plans laid out, consider the way you will walk through the entire space. Eliminate unnecessary wasted areas; remember you are paying by the square foot.

- **Doors that connect awkwardly.** Examples are a bathroom that connects to a kitchen, or a bath that can only be reached by guests walking through a bedroom. It is also awkward when the back door can be seen directly in front of the front door, or when the front door seems to lead to the stairs instead of the living area.

- **Master bedroom on the front of the house.** If the house is on a street with any traffic, the master bedroom will seem noisy.

- **Building the biggest house in the neighborhood.** This can make your house harder to sell.

Basements: How Important are They?

Whether to put in a basement is an important consideration to which you should give much thought. Your decision will be based partly on the area in which you live. A good, dry basement can be a bonus, but a wet basement will make you wish you had never had one. A basement that allows enough water to enter the home can eventually cause the foundation to be destroyed.

If you want a basement, it is best to avoid flat lots. A sloped lot will allow you to create good drainage and avoid potential problems with dampness or mold. Sloped lots also allow for walkout basements, which have both natural light and the possibility of ventilation.

Because basements can be problematic, consider whether you need one. If the water table is high or the land does not drain well, flooding can be an annual event. A damp basement may require a dehumidifier, professional waterproofing, and constant work.

On the positive side, a basement is fairly inexpensive to build. It gives extra storage space and a place to put the heating system. It can also provide space for a home office, workshop, or getaway den for the teenagers.

Designing to Save Money

You can cut costs at the design stage. If, after you have the basic plan you realize there will be cost overruns, consider some of the following ways of saving money.

- **A rectangular floor plan.** Use a floor plan that is as near a rectangular shape as you can get it. This is the most efficient plan. It allows for a simple layout and a simple roofline.
- **A smaller footprint.** Bumping out bay windows and breakfast areas several feet break up the straight lines of the house and add architectural interest. They give you more sure footage without costing you anything for an enlarged footprint. The foundation will be smaller, so the cost is lower.
- **Building close to the street.** Consider building the house closer to the street. Driveways are expensive. You can save thousands on concrete just by moving the house on the site. Be sure to allow for the proper setbacks.
- **Eliminate the basement.** Consider whether you really need that basement. Basements can add 5 to 7 percent to the cost of a home.
- **Add the extras later.** Consider putting off some of the extras you want. By only rough finishing the basement or upstairs, you can save money now, and then finish it as needed.
- **Lower the ceilings.** If you have chosen 9-, 10-, or 12-foot ceilings, you can save money on drywall, paint, trim, and heating costs by lowering the ceilings.
- **Combine rooms.** Each interior wall costs money. If you can reduce the need for separation, you can save money and give the double room an airy feel.

- **Choose factory built homes, modular homes, or "panelized" construction.** Panelized buildings are somewhere between modular and stick built. Pre-constructed wall panels, with the windows, doors, siding, and insulation already installed, are shipped on two trucks, unloaded, and assembled. These homes are constructed in only a few days. When the construction is finalized, observers cannot tell that it was a panelized house. Kit homes that are not pre-constructed can also offer cost savings. When you purchase a kit home, you know the exact price of the materials.

Seven Steps to the Perfect Home Design

1. Make a list of all your wants and needs.
2. Select your favorite three home designs as the final contenders, so you have options for fitting on the lot.
3. Include the possibility of future expansions to your home, even if they are as minor as finishing a basement or attic.
4. Make sure you have room to move around even after you have put in your furnishings.
5. Make sure kitchen traffic flows well from the sink to the stove to the refrigerator, and there is enough counter space to work easily.
6. Consider putting the laundry room on the same floor as your bedrooms. It makes sense to not have to carry loads of laundry up and down the stairs.
7. If you have special hobbies like sewing, painting, scrapbooking, or playing music, consider incorporating your own playroom into the plan. It can be custom designed to fit your hobby, and the rest of your family will appreciate the clutter or noise being contained. One artist said her dream house would have a separate room for each type of craft.

Chapter 2 Checklist

☐	Look at available house plans.
☐	Consider drawing your own plans.
☐	Consider the use of an architect or designer to draw or modify your plans.
☐	Determine which of the above approaches you will use.
☐	Consider to what degree you want to use green materials and techniques and include that in your plans.
☐	Determine the placement of the house on the lot.
☐	Review your plans to make sure they include:
☐	A plat plan
☐	Spec sheets
☐	Floor plan
☐	Foundation plan
☐	Detail sheets
☐	Elevations
☐	Mechanical, electrical, and plumbing layouts (possibly by others)
☐	Check the floor plan for:
☐	Traffic patterns (door locations)
☐	Window locations
☐	Storage areas and closets
☐	Convenience for daily activities
☐	Determine whether to have a basement.
☐	Consider money-saving design ideas.
☐	Review the seven steps to the perfect house design.

CHAPTER 3

Pre-Construction Activities

Many preparations must be completed before you begin construction. You have already done some of them: you started your notebook, began reading about construction, and created your house plans. Now it is time to move on to the pre-construction activities, like estimating costs and gathering tools. At this point, the to-do list may seem endless. This process must be accomplished one step at a time.

The process should not be rushed. Most contractors recommend that first-time builders spend at least as much time on planning as they do on building. That means if you are going to build your house in six months, you should plan for six months first.

Before beginning these activities, walk through your floor plan one more time. Make sure that the plans are exactly the way you want them. Making changes on paper is easy — but changes made, for example, to the windows after they have already been installed will cost you money, time, and labor.

The Steps of Construction

This chapter covers the first major step, pre-construction. After that the other steps occur in this approximate order:

- Excavation
- Foundation
- Framing
- Roofing/gutters
- Plumbing
- HVAC
- Electrical
- Masonry
- Siding
- Insulation
- Drywall
- Trim
- Painting, staining, and wall covering
- Cabinetry and counters
- Flooring
- Landscaping

Draw Up the Construction Schedule and Flow Charts

A written schedule with deadlines will encourage most people to get the project done on time. According to the National Association of Home Builders, the average owner-builder takes nine and a half months to complete a home. But with an organized, written plan, a reasonable goal could be six months. The best way to start is by writing out a timeline of how things will proceed.

The Many Steps Involved in Construction

Construction involves many processes, and each one is dependent on one or more of the others. Activities may take more or less time than you expect; there may be scheduling delays and weather delays. Therefore, the entire plan has to be somewhat flexible.

It is also important to get the building permit, insurance, and other required preliminaries before creating your schedule. There is nothing more frustrating than delaying construction because there is no permit yet.

On the other hand, construction loans have pre-defined longevity. Some are four months, while others may be for six months. So your timeline may be influenced by the loan. Know your loan terms before creating your timeline.

How to Form a Construction Schedule

Lay out your timeline carefully. To begin creating a timeline, ask subs how much time they need for the job, and how much notice they need when you are ready for them. Ask suppliers how much lead time they need to get your materials, and include those details on the timeline. The worst feeling in the world is holding up your own construction because you forgot to order something crucial.

After you think through your schedule and write it out, have your site supervisor look it over. If you are not hiring a supervisor, consider asking a sub or a supplier to review your written schedule. Either will be happy to offer suggestions for running the project more smoothly.

It is easiest to use a computer program like Microsoft Excel to create a construction timeline. Be sure that every single item is listed, and that the list is in the correct order.

When you are finished, step back and look at it. Does the flow make sense? If it does, go to each sub and get a firm commitment in writing. This should include a start date, an estimated duration for the project, and a finish date. Make adjustments to your timeline if necessary.

How to Form a Construction Flow Chart

A construction flow chart is exactly like the timeline, except it is a bar graph, or picture, instead of words. In a way, the flow chart is easier to "read" because you simply glance at the colored blocks.

Here is a sample:

FLOW CHART FOR ALLEN ADDITION

	Week:	1	2	3	4	5	6	7	8	9	10
	Date:	11/10	11/17	11/24	12/1	12/8	12/15	12/22	1/04	1/11	1/18
PHASE	**NAME**										
Site setup	self										
Deconstruction	self, one crew										
Site Earthwork	AA Excavating										
Foundation Concrete	AA Excavating										
Frame	DeLong Custom Framing										
Windows/Doors	self										
Stucco	Alberston										
Mechanical	Superior										
Plumbing	McDaniels										
Electrical	none yet										
Insulation	Carter										
Drywall	Blevins										
Cabinetry	Zibbers										
Finish Carpentry	self										
Paint	self										

Cost Estimation

Estimating costs is one of the most time-consuming tasks you will perform. It is also one of the most important, if you are constrained by a construction loan or spending limit. Time spent on cost estimation, and time spent weekly studying expenditures, will prevent cost overruns.

One important source of costs will be your subs. Although you can get estimates from suppliers, the subs will have a better idea of what everything will cost.

The construction plans you purchased may have come with a "materials list." If you have to pay extra for it, do not bother; often these lists are not reliable. If you do use it, only treat it as guide — run your own estimates instead.

When figuring your estimates, use the checklist at the end of this chapter as a guide. Take the numbers from the list and enter each onto a purchase order form for your various suppliers. Remember to control costs from the beginning; you will not save 10-35 percent of the cost of having someone else build your home if you do not control your spending.

The following tips will help you create estimates:

Excavation is charged by the hour. Larger equipment costs more to rent, but can perform the work in a short time. Clearing a half acre of land may take from 1 to 4 hours; digging the foundation might take around 3 hours.

Concrete estimates are figured on 4-in. slabs. A cubic yard of concrete is required for 81 sq. ft. of slab. Footings are twice as wide as the walls, and there will be a local code requirement that determines the total dimensions. After checking the code, calculate the cubic yards of concrete needed, and include 3 extra feet of depth for every pier hole, plus at least 10 percent for waste. If the depth is 12 in. and the width is 8 in., you will need 0.025 cubic yds. of concrete for every lineal foot. These figures are for smooth finish work — not driveways or sidewalks.

Roofing from scratch is harder to calculate than other types of roofing. The plans should help you to calculate the rafter length and the roof area. The pitch of the roof is described as inches per foot of distance. In other words, a 10-in.

slope rises 10 inches for every foot of horizontal distance. Once you have figured the square footage of the roof, add 1½ sq. ft. for every foot of eaves, ridge, or valley. Divide the total by 100 to get the number of "squares." Roofing shingles are sold in squares, which is the number of shingles that will cover 100 sq. ft. of roof. Roofing felt goes underneath the shingles and is available in 500-sq.-ft. rolls. Divide the square footage of the roof by 500, and then add 20 percent for waste to determine how many rolls you need.

Use bids from subs. If you are going to use subs to do the work, always allow them to give a bid to complete your estimate. It takes a lot of time calculating your costs, and they may do it faster and more accurately. Always use the bid price over your own estimate, making sure that you agree on the square footage of the area to be covered.

Have a mortgage survey done. Surveys are required twice by the lender. The first is a vacant land survey and the second is a mortgage survey. Even if you do not have a lender, it is a good idea to have the mortgage survey done so that you have a record of the site, setback requirements, and where the house is located in relation to the property lines. This can prevent many problems later on. I remember one owner-builder who placed his house 15 feet over the setback. Because he did not have a mortgage, no one caught the mistake until the house was under contract, and the deal fell through because of the error.

Include the cost of a water supply. Water tap-in fees can be obtained from the municipality. If there is going to be a well, get a bid from a drilling firm. Be sure that their bid includes a maximum and a price-per-linear-foot.

Chapter 3 Checklist

☐	Walk through the floor plan one last time and make any desired changes now.
☐	Determine the construction timeline and create the construction flow chart.
☐	Develop a materials list.
☐	Estimate the cost of the project.

CHAPTER 4

Financing

Getting a construction loan is a simple process, but just like with any other type of home loan, it pays to shop around. The good news is that the process has become easier and the costs have become lower in recent years. Additionally, there are more choices in loans now than there were ten years ago.

You must be business-like when approaching a lender. If you put yourself on the lender's side of the table for a moment, you can see that the lender is taking on risk to loan you money — so you must convince him that the loan is not much of a risk for him. The more you can convince him that you are a low risk, the more likely you are to get a loan. Convince the lender by preparing adequately for the application. Get prequalified for the loan, and present a complete, well-documented application package.

The risk the lender is taking falls into several categories. The first is the normal risk that a lender takes on every time he makes a loan. Are you financially sound? Are you likely to pay the loan payments? Your credit score and financial qualifications will make the lender more confident. If he sees a strong financial history, he will feel that you are likely to repay the loan.

In addition to the financial responsibility, the building project itself is a risk to the lender, especially if you have never built before. Will the home meet code? Will it resell? Can you keep the project within budget? Can you get it done in a reasonable amount of time?

These are all valid questions. Lenders know that custom homes vacuum up money, and that most of the time they go over budget. To satisfy the lender, you must be willing to take some extra steps and approach him with either a proposal or a portfolio that answers all the questions he might ask. You need to convince him that your loan will be no risk. Show him that you will not leave any of the outcomes to chance.

Prequalification

The best way to be sure you will qualify for the funds before you spend a great deal of time and money on plans is to get prequalified for the loan. It is easy to complete the prequalification process, and it is free. When you are finished, you will know for sure that the loan funds you intend to request are within your grasp.

"Get prequalified" means, "Find out how much you can borrow." By taking this extra step, you will know right away whether you can afford the 3,500 sq. ft. home you were contemplating. Prequalification is fairly informal, but it will give you the confidence you need to prepare for the "real" application.

The forms required by lenders vary, but you will need to complete an application, a description of materials, and a construction cost breakdown. The application requires your employer information, assets and liabilities (including account numbers, balances, and the addresses of your banks), and income information

(W-2s or tax returns). The lender can quickly tell you how much you will be able to borrow.

Shop Around

After getting prequalified, look for the loan and the lender you will use for your construction process. It takes time to find a lender, and every lender has different requirements, so start at least several weeks ahead of the date that you want to begin construction. Check to find out what the lender you select wants to see in your application. It pays to look around — especially since, after reading this book, you will be the most prepared applicant your lender has ever met. You will be considered a good candidate, so good that the lender will even want to negotiate terms for you.

Your Shopping List

The questions you ask when you get a construction loan are almost the same as for a mortgage loan. You will need to know how much the origination fee is, what the interest rate is, and whether the rate is adjustable or fixed. Can you lock in the rate? For how long? What will it cost? Is a down payment required? How much? What will your other costs be?

Loans and Points

A construction loan has two types of fees: the interest rate, which is calculated over the life of the loan; and the points, which are upfront fees that can offset the interest rate. Just as you negotiated the price of your land, you can negotiate interest and points somewhat. You might be able to "buy down" your interest rate by paying a little more in points.

A point is equal to 1 percent of the loan amount. There are virtually no construction loans on the market for 0 points. You should expect conventional lenders to offer a loan with 2 or 3 points. If you offer to pay more points — say, 3 instead of 2 — you will have an interest reduction of 1 point. The loan officer can help you determine which situation is best for you.

Which Loan Is Better?

When you compare two loans, the terms are often different, making it seem impossible to compare them. You can find online calculators that will compare the terms of two loans by typing "compare two loans" into a search engine.

Let us look at another way to compare loans. Imagine you are borrowing $200,000. One bank offers you 2 points and 5 percent; the other wants 3 points, but the rate is only 3.75 percent. To compare them, figure how much you will pay in points, then multiply the loan by .6 (for the construction portion of the loan) and calculate the interest.

Loan 1:

200,000X.02=4,000.0 200,000X.6=120,000.0 120,000X 0.05=6,000.0
4,000 + 6,000=10,000 Loan 1

Loan 2:

200,000X.03=6,000.0 120,000X0.0375=4,500.0
6,000 + 4,500=10,500 Loan 2

Draws

Construction loans differ from other loans you might have gotten in the past. When you take out a construction loan, you are able to draw (pay yourself the funds) at periodic intervals. You will need to know what the draw procedures are. Is an inspection required? Is notice required? Is there a fee? How often are you able to get a draw? What is the lender's procedure for releasing the lien?

Types of Loans

Many borrowers wonder whether they are shopping for the loan and its terms, or shopping for a lender. In a sense, you are doing both. Ideally, the lender who gives you the best service and listens to your needs is also the one who offers the best loan package. At any rate, you will be better off if you shop around for the loan.

Should you use the lender you used the last time you purchased a home? Probably not, if you did not build it yourself. Construction loans are different from conventional home purchase loans. If you can, find an experienced construction lender. Local banks may not have a construction loan department. Test their loan officers by asking them questions about the differences between construction loans and other kinds of loans. Ask why the fees vary, and why certain forms are required, and ask whether the draw reimbursement is preferable to a voucher system. You will soon be able to determine whether the loan officer understands construction loans.

There are several kinds of loans. Look for a lender or mortgage broker who will listen to your needs, and present you with honest answers rather than trying to sell you on the loan package of the week.

Land Loan

Your first loan will be for the lot you plan to build on. After negotiating the price with the seller and getting a sales agreement in writing, consider the available financing options. If the seller will act as the lender, referred to as "carrying back paper," this may help you out over the short term, but when the bank funds the construction loan they will require that the seller be paid off. Normally, seller financing does not give you a better deal, and in many cases, sellers charge as much as ½ to 1 point more in interest than the banks. A seller might offer to carry the loan to speed up the sale, or to put off paying the taxes on the sale of the property.

Occasionally, a seller offers financing because he knows banks will not lend money on the property. Sometimes this can be good, and sometimes it is because the seller is being deceptive. I ran into this problem myself a few years ago. A seller had 32 acres of beautiful land and offered to carry the loan. I know that lenders are wary of self-employed "people like me," so I found the offer appealing. I signed a contract to purchase. A few days later I visited the property again, and this time a neighbor stopped me. He said he owned the road, and would move the road to another side of his own land so that I could not get to "my" property. Further investigation revealed that the neighbor was right; the acreage was landlocked. I spoke with county officials, who said that most people in this especially rural area would simply sell part of their land to give the neighbor access to the road, but in this case no one wanted to. I confronted the seller, who admitted he knew this. He refunded my money and canceled the contract — but this could have been a costly mistake. He could have held me to the purchase contract, and I would have owned a tract of land I could not build on, or even get to. If you believe the seller is too eager or he insists on financing the land, watch out!

There may be other reasons property cannot be financed. If the property is considered very large — over 20 acres — you might have to look harder to find a lender. If it has no electricity, or it is already divided into multiple parcels, or if there are buildings on it, you may not be able to obtain conventional lending.

The length of the loan and the interest rate are two of the most important ingredients of a land loan. The majority of lot loans cover a period of less than five years. If you believe you will get to the construction loan phase within two or three years, this should be fine. But if you feel that you will take a long time to design the house, or the local planning department is notorious for delays, it might be better to search for a ten-year land loan. Whether you choose a fixed or an adjustable rate may have less significance, depending on how soon you are going to start building. If you are getting the construction loan fairly soon, it will pay off the land loan anyway.

Many buyers mistakenly believe that they must pay off the land loan before obtaining a construction loan. This is not quite the case. As mentioned in Chapter 2, you can do this to use the equity you have in the land as leverage. Years ago, as a standard practice, owner-builders bought the land, paid it off, and used it as collateral for the construction loan. You can do the same if you want to, but these days a construction loan replaces the existing land loan when the time comes. Instead of paying off the land, you can use that money to pay the architect, designer, or contractor, and perhaps to pay for the many permits that are required. In addition, the interest you are paying on the land loan, although it is not much, is tax deductible. It is better to keep the land loan and leave the cash in your pocket.

Construction Only Loan

A construction only loan is the most common and the oldest type of construction loan. It is a short-term loan that only covers the construction period, which will be agreed on with your bank as some period from 6 to 18 months. As soon as construction is complete, you must arrange for another loan as the "permanent" financing. This not only means you must apply and qualify for another loan; you must pay all the loan fees and costs for the new loan.

Many things can change while you are building the home. Will it be finished at the end of the construction loan term? Will the market or interest rate change significantly? Will the house qualify?

Banks are not sympathetic when the 12 months are up and an owner-builder is unable to afford the new loan. For this reason, it is better to get a permanent construction loan, also referred to as a "single close" or a "one-time close."

No-Income-Qualifier

You may hear these loans called names such as "No Doc" or "EZ Qualifier." They are marketed as requiring less documentation than other types of loans. For example, you might be able to state your income and assets without providing verification of them. The idea appeals to people who do not want to disclose their income for some reason, as well as to people who are self-employed and do not have conventional documentation. These loans might require better credit and, in the end, may cost more than the documented methods of borrowing money.

You will need to provide a copy of the deed or the purchase contract, plus you will need the plans, materials description, and the cost breakdown. You might also be required to provide documentation of assets.

Construction-Permanent or Permanent Loan

These single-close loans were created in the 1990s, when banks finally realized that owner-builders were among their best customers. After all, after you put all that work into building your home, are you going to lose it to default if you can help it?

The permanent construction loan rolls over into a long-term mortgage after construction is complete. This means you do not have to go through another qualification process or pay to have the property re-appraised. You do not have to pay for a second closing. Indy Mac even offers a package that does not require payments during construction. Instead, you create an interest reserve used to pay the interest that is accruing on the loan.

Paperwork the Bank Will Need

A residential loan application requires an incredible amount of paperwork. Fortunately, the lender will help you fill it out. If the permanent loan follows a construction loan from a different lender, you will need all the paperwork from your construction loan. Expect to provide:

- A complete cost breakdown (sample below)
- A builder's statement if you are working with one
- A copy of the builder's license, if it is required in your state
- Copies of insurance policies, including builder's risk, flood insurance, liability, and worker's compensation (if appropriate)

Sample cost breakdown for bank loan application:

LOAN INFORMATION

BORROWER NAME:

PHONE #:

PROPERTY ADDRESS:

LINE ITEM DESCRIPTION	TOTAL PROJECT COSTS	BORROWER PREPAID COSTS	CHANGES TO BUDGET	REMAINING FUNDS
A. PRE CONSTRUCTION COSTS:				
1 Architect, Engineering & Soils Study Fees				
2 Design Review/ Plan Check Fees				

LINE ITEM DESCRIPTION	TOTAL PROJECT COSTS	BORROWER PREPAID COSTS	CHANGES TO BUDGET	REMAINING FUNDS
3 Permits - City/ County				
4 Utility Connection Fees				
5 School/Park/ Misc. Taxes				
6 Project Bonds				
TOTAL PRE-CONSTRUCTION COSTS				
B. GENERAL REQUIREMENTS				
7 Temporary Utilities & Facilities				
8 Special Inspections/ Testing-Geo-tech, Structural				
9 Job Security				
10 Equipment Rental				
11 Jobsite overhead				
12 Project Management/ Supervision				
13 General Contractor's office overhead/profit.				
14 State Sales Tax (where applicable)				
15 Builder Contingency				
SUB-TOTAL GENERAL REQUIREMENTS				
C. SITE PREPARATION				
16 Demolition				
17 Clearing/Stakeout				

LINE ITEM DESCRIPTION	TOTAL PROJECT COSTS	BORROWER PREPAID COSTS	CHANGES TO BUDGET	REMAINING FUNDS
18 Rough grading/ shoring/excavation/fill				
19 Site retaining walls/ waterproofing/backfill				
20 Site drainage				
21 Private septic system				
22 Domestic Water well				
23 Pump house & Pressure water system				
24 Environmental				
25 Off-site improvements				
SUB-TOTAL SITE PREPARATION				
D. FOUNDATION COMPLETE WITH+A15 FOUNDATION ENDORSEMENT				
26 Embedded hardware				
27 Ground Plumbing				
28 Ground Mechanical				
29 Ground Electrical				
30 Underground utilities				
31 Foundation & Building retaining walls poured				
32 Concrete slab poured-house, garage				
SUB-TOTAL FOUNDATION COMPLETE				
E. BUILDING ROUGH-IN COMPLETION				
33 Structural masonry				

LINE ITEM DESCRIPTION	TOTAL PROJECT COSTS	BORROWER PREPAID COSTS	CHANGES TO BUDGET	REMAINING FUNDS
34 Rough framing materials				
35 Structural steel				
36 Modular or Sectional Mfg. Home				
37 Package/Kit Home				
38 A51 Trusses/ components				
39 Rough framing labor				
40 Lightweight concrete interior floors				
41 Plumbing top-out				
42 Rough heating, ventilation, air conditioning				
43 Rough electrical				
44 Fire protection - sprinklers				
45 Fireplaces and Flues				
46 Security & Communications pre-wiring				
SUB-TOTAL BUILDING ROUGH-IN COMPLETION				
F. EXTERIOR WEATHER-TIGHT				
47 Waterproofing decks, etc.				
48 sheet metal, Gutters, downspouts,				
49 Roof covering				
50 Windows				

LINE ITEM DESCRIPTION	TOTAL PROJECT COSTS	BORROWER PREPAID COSTS	CHANGES TO BUDGET	REMAINING FUNDS
51 Exterior doors				
52 Skylights				
53 Glazing				
54 Exterior siding				
55 Exterior trim				
56 Stucco				
57 Masonry veneer				
58 Ornamental Iron				
59 Garage Doors				
60 Exterior painting				
SUB-TOTAL EXTERIOR+A55				
G. DRYWALL/FINISH CARPENTRY				
61 Insulation				
62 Drywall/Plaster				
63 Interior stairways				
64 Cabinetry				
65 Finish Materials/ Millwork				
66 Interior Doors				
67 Finish Hardware				
68 Finish Carpentry Labor				
SUB-TOTAL DRYWALL/FINISH CARPENTRY				
H. BUILDING COMPLETION/FINAL INSPECTION				
69 Countertops				
70 Tub/shower/ enclosures				
71 Interior painting/ Wall Coverings				

LINE ITEM DESCRIPTION	TOTAL PROJECT COSTS	BORROWER PREPAID COSTS	CHANGES TO BUDGET	REMAINING FUNDS
72 Hard surface finish flooring				
73 Carpeting				
74 Built-in Appliances				
75 Special Equipment				
76 Security system				
77 Intercom				
78 Built-in Vacuum Cleaner				
79 Finish Plumbing				
80 Plumbing Fixtures				
81 Finish Electrical				
82 Lighting Fixtures				
83 Finish Heating, Ventilating, Air Cond.				
84 Solar Backup				
85 Bath Accessories				
86 Tub, Shower Doors, Mirrors				
87 Finish Grading				
88 Pool/Spa				
89 Hardscape (Driveway, Walkways, Steps)				
90 Landscaping				
91 Irrigation System				
92 Fencing and+A90 Gates				
93 Touch-up/ Final Cleaning				
SUB-TOTAL BUILDING COMPLETION				

LINE ITEM DESCRIPTION	TOTAL PROJECT COSTS	BORROWER PREPAID COSTS	CHANGES TO BUDGET	REMAINING FUNDS
TOTAL CONSTRUCTION COSTS				
TOTAL LINE ITEM COST BREAKDOWN				

When To Apply

You may be wondering when to apply for your construction loan. After all, the loan documentation, like your credit report and appraisal, will expire after 90 days. So if you apply and then do not fund the loan within 90 days, you will have to pay more fees and reapply. On the other hand, if you have to wait around on the financing, you could waste valuable construction time, or lose one or more of the subs you had lined up.

Most building departments take 1-2 months to issue permits, so if you apply for the loan about 2 months before you believe you will break ground, you will have the funding at exactly the right time.

If you have already started building before you apply for the loan, you can get it funded easily. The bank may require that you provide them with title insurance and a deed of trust.

Truth in Lending

Because of the Federal Truth in Lending Act, people who want to borrow money to buy or build real estate have certain rights. Article Z of this Act requires that the lender disclose to you the details of the loan he or she is offering you, including:

- The interest rate
- The terms
- The extra costs
- Variable rate features in the APR

If the lender does not disclose the APR of the mortgage loan to you, then you are able to receive a refund of your application fee. You will normally receive all the disclosures in writing when you make the loan application. If any of the terms change before closing, and you decide not to proceed with the loan, the lender must return all fees to you.

After closing, you have three days to change your mind. If you make a request in writing to the lender within three days, all your fees will be refunded.

Obstacles

There can be obstacles when applying for a construction loan. You can get around most of these obstacles if you are prepared for them ahead of time. The best way to prepare is to check and double-check on financing, the house plans, and the process. Be prepared to change gears and be as flexible as possible throughout the project.

Credit

Your credit is one of the first things that can get in the way of your construction project. Credit is based on a scoring system called the FICO score. "Poor" credit is considered to be about 300; average is 620; and good credit is 660 or better. The score goes up as high as 850.

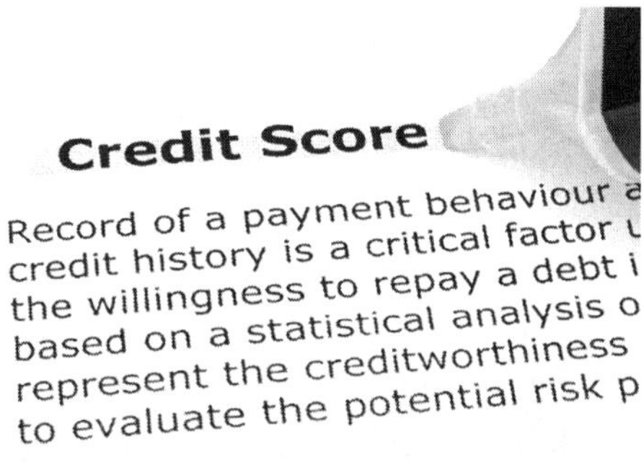

Credit scoring is based on your pattern of making payments (35 percent), how much money you owe (30 percent) and the amount of available credit. Lenders base their loan approval, and possibly the interest rate, on your credit score. If your credit score is below 620, you may have problems getting a loan through conventional banks — just as you would with any other type of loan. If you know your credit score is low, approach non-conventional lenders.

The cost of bad credit is high when you are considering permanent loans. If your credit score is 650 and the rate offered to you is 1 point higher than that offered to others with better credit, you could end up paying $21,000 more over the life of a $170,000 loan.

Credit reports often contain inaccuracies, and sometimes they contain information that does not belong to you. As soon as you begin thinking about building a home, get a copy of your credit report. These are free once a year from each of the three credit reporting agencies listed below. You can also request a free annual report by visiting www.annualcreditreport.com or by phoning 1-877-322-8228.

Once you have a copy of your report, study it carefully.

If there are problems or false information, contact the agency right away. You can do this by mail or through the Internet. Give the agency supporting documentation of the item in dispute. The complaint will be investigated within 30 days; complaints are resolved quickly.

You can get a copy for free from any of the "Big 3" credit agencies:

Equifax

PO Box 740241
Atlanta, GA 30374-0241
800-685-1111

Experian

P.O. Box 2002
Allen, TX 75013
888-397-3742

TransUnion

P.O. Box 390
Springfield, PA 19064
800-916-8800

Unexpected Fees

When building a home, you will feel as if the array of fees is endless. The best way to avoid surprises is to ask and re-ask the lender, the local government agency, and your subs and suppliers about fees. There will still be unexpected charges, so it is wise to keep strong cash reserves.

One surprise, especially if you are already familiar with purchasing real estate, is the appraisal. Construction loan appraisals require more time and more cost than the average purchase loan appraisal, so the extra expense is passed on to you. If you are borrowing a lot of money, you can expect the cost of appraisal to be especially high — perhaps double what you anticipated. Additionally, an appraisal only lasts 90 days and you will be charged $150-$250 for having it updated if you need it for longer. And if you change lenders, you will pay a fee to have the appraisal retyped, if the appraiser agrees to it. Often they do not, and you will have to order a completely new appraisal.

"Junk" fees are another surprise. These are the small fees charged by various parties in your construction loan. Lenders will break this down for you in writing. Despite the name, most of the costs are necessary. There may be inspection fees, administration fees, wire transfers or messenger fees, processing, recording, and certifications. Get a good faith estimate in writing up front and examine the list carefully — but most of the fees are for services that must be paid for to process the loan.

Passing Muster

Often the loan committee at the bank will need to see qualifications for building a house that you do not have. In this case you could call around and find a general contractor who will provide a copy of his license for the files. This seems to satisfy the requirement as far as the lender is concerned. You might try to find a semi-retired contractor to do this — or even offer a fee. If the bank asks that the contractor sign an agreement to perform services, let him select an hourly rate for consulting. Even if no money changes hands, this seems to be satisfactory. This is just a formality – the bank wants to have evidence that the money it is lending is paying for a house that will have value in the end.

Being too Conservative on the Time Frame

If you get a construction-only loan, it will be offered for somewhere from 6 months to 18 months. Many people mistakenly choose the short term, only to be surprised later. The shorter-term loans seem attractive, since they carry a better interest rate. But if you extend past the loan term, you could end up paying large penalties, as high as 0.5 percent of the loan amount (not the interest — the total loan). That can end up being quite costly.

Many people set a goal of 6 months, and some use a goal of 4 months to completion. In theory, there is no reason you cannot complete the construction of a house in that amount of time. In reality, you do not know the system yet. You do not know the inspectors, the government entities, nor the subs. There is a good chance that there could be delays during the construction process.

To avoid the stress and extra expense, try to calculate the length of time you believe the home will be under construction, then double it and use that as the loan term. This should allow enough of a cushion to offset delays, weather problems, losing subs, and other setbacks. The extra percentage in interest will be very little compared to the tension people feel when the loan term is nearly up.

Budget will not Work with your Loan Abilities

Not talking with lenders until after the permit process can get you into a lot of trouble. All lenders have certain guidelines, and if your budget does not fit, you might not qualify for a loan. This can delay the project for years. I cannot emphasize this enough: find out what you can afford to build first, then create a plan.

Checklist for Construction Loan Documents

☐	Application form
☐	Consent form
☐	Signed disclosure
☐	Most recent pay stubs
☐	2 years' W-2s or last 2 years' tax returns (if self-employed)
☐	Rental agreements, if you own rental properties
☐	Other asset records
☐	3 months of bank statements on all accounts
☐	Retirement account statements
☐	Land contract or deed
☐	Loan documents (for land with loan)
☐	Survey
☐	Proof of regulatory compliance
☐	Plans (3 sets)
☐	Specifications
☐	Cost breakdown
☐	Title insurance
☐	Construction insurance
☐	Architect information
☐	Permits
☐	Builder's contract
EXTRA DOCUMENTS THAT WILL MAKE YOU STAND OUT FROM THE CROWD	
☐	Copies of subcontractor and supplier bids
☐	Copy of your project timeline
☐	The bank's own lien release form, signed by subs (this is a form subcontractors sign saying they have been paid and will not put a lien on the property)
☐	Resumes of contractor or major subs you are using

Chapter 4 Checklist

☐	Obtain your credit report.
☐	Prequalify for the loan.
☐	Investigate financing.
☐	Understand points versus interest rate and how to calculate the best deal for you.
☐	Understand different loan types and draws.
☐	Shop around for a loan.
☐	Select lender.
☐	Complete checklist for construction loan types and draws.
☐	Complete loan application.
☐	Apply for loan.
☐	Review potential problems.
☐	Prepare for unexpected fees.
☐	Review budget.
☐	Timeframe for construction.

CHAPTER 5

Financing for Remodeling

Financing a Remodeling Project

If you are preparing for an extensive renovation or addition project, you may be wondering where you will get the money. Most people do not realize how easy it is to get financing for a home upgrade, especially if you have built up equity in the house.

The easiest and quickest way to get money to finance your project is through a second mortgage on the home. This can be a home equity loan or a Home Equity Line Of Credit (HELOC). The difference between the two is that with a loan you have a closing and receive all the money at once; with a line of credit you can make draws and take out the money as you need it, so you save a little in interest if you do not draw out the money until the moment it is actually going to be used.

Your home serves as collateral for both types of loans. You will be approved for a loan limit based on the home's appraised value and the equity that you have in it. For example, if your home is worth $100,000, the banks will probably allow a credit limit of 75 percent of that, or $75,000. But if you still owe $35,000 on the home, you only have $40,000 in equity built up in the property. So the lender will loan you up to that amount — $40,000.

Many of the HELOCs are good for a specific period of time, like five years. During that time you can draw on the money up to the agreed-upon limit. At the end of the time, you cannot borrow additional funds unless you renew the loan. Some plans have a minimum draw or a minimum outstanding amount.

Most HELOCs offer a variable interest rate. That means it changes with the prime rate and will be offered as that index plus a margin, like 2 percent. Variable rate plans must have a cap, or limit, on how much the interest rate can increase over the life of the loan. Be sure to find out how high your interest rate can go. Also find out whether the loan will allow you to convert the variable rate to a fixed rate should you decide to. The interest on home equity loans is tax deductible — it depends on your personal financial situation, so check with your tax advisor to be sure.

The Cost of a HELOC

The costs for home equity loans include an application fee, an appraisal fee, and some upfront charges like 1 or more points. There will also be the normal closing costs: attorney fees, mortgage preparation, taxes, property insurance, and title insurance. There might also be transaction fees and additional fees each time you draw money. These costs should all be revealed to you in writing when you apply for a loan.

The Cost of a Second Mortgage

Compared to a line of credit, a second mortgage loan might seem like a better deal. It will give you a fixed amount of money, and it can be paid over a fixed time period. If you choose a fixed rate loan, all the payments will be equal. When comparing the loans, be sure to compare the closing costs and the interest rates — but the APR for a line of credit is figured differently. It is based on the periodic interest rate, whereas the APR for traditional mortgages is figured on the interest plus points and other finance charges.

Applying for the Loan

Just like the construction loan, a home equity loan will require a breakdown of labor and materials costs. Be sure to include all your permit fees and equipment rental, then add a cushion of at least 20 to 30 percent for unexpected expenses. Remodeling tends to have more cost overruns, in part because you are not aware of problems that may be uncovered as you begin work. Hidden structural defects, electrical wiring that needs to be replaced, or other surprises may cause more money as well as more time spent to be spent on the project.

The amount of money you can borrow will depend on the value of your home, the loan-to-value ratio, your income, and your credit rating. The house payment plus your other debt should fall below 36 percent of your gross monthly income to qualify for the loan. For credit that is less than perfect, you may need to pay points to get the loan.

Special Loan Programs

A Federal Housing Authority (FHA)-backed loan, called an FHA 203(k) mortgage, allows you to refinance your home and roll remodeling costs into a new loan. The loan is based on the estimated worth of the house after you have completed the improvements. The loan limits will vary by county; they tend to be somewhat low, but if you qualify you will seem to have more equity, and therefore the amount you can borrow will be higher than with some other loans.

Another kind of mortgage is an Energy Efficient Mortgage, or EEM. This kind of loan can increase your debt-to-income ratio by two percent. Your home will have to meet stringent energy efficiency standards to qualify; check with Fannie Mae for details.

B and C Loans

These loans work for people who do not have good credit, or who do not fit the normal employment criteria. Lenders push them for debt consolidation, but they are available almost everywhere. They are non-conventional loans, so the interest and fees may be higher. They are also difficult to compare because the requirements and terms vary.

Other Collateral

If you are borrowing to improve your home, and you have assets other than your house, you might use a different type of loan. Stocks, bonds, savings, retirement accounts, and CDs can be used as loan collateral. There are not the closing costs associated a second mortgage or home equity line of credit, but the interest is not tax-deductible. The interest rate might be low enough to make the loan worthwhile.

Obtaining financing for a remodeling project is much easier than getting financing to build a new home. It is also much faster. Be sure that you check into all the options that are available to you, and that you understand all the terms of the contract before you sign.

Chapter 5 Checklist

☐	Investigate costs and availability of HELOC versus second mortgage loans.
☐	Investigate other available types of loans.
☐	Investigate government loans such as FHA and EEM.
☐	Investigate B and C loans.
☐	Investigate collateral loans (other than your home).
☐	Investigate private loans.
☐	Determine the best type of loan for your situation.

CHAPTER 6

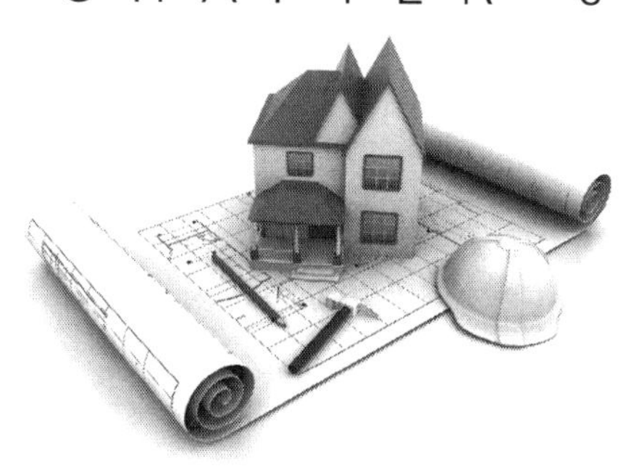

Insurance

Obtaining Essential Insurance Information

Several kinds of insurance will be necessary throughout the purchase and building process. Whether you are working with a lender who requires it, or financing your project 100 percent yourself, you need protection. Not all builders' policies are covered by every agency, so give yourself plenty of time to find the kind of insurance you need. Following is a list of some of the more common kinds of insurance that will be necessary.

Title Insurance

Title insurance is a policy that you obtain when you purchase the land. The seller should insure that the property has "clear title." That means that it is free of legal defects, liens, and encumbrances unless they are listed in the policy. Most property in the United States has gone through ownership changes, and there could be a weak link somewhere along the way. For example, there could be liens, unpaid taxes, or fraud. Title insurance covers you against any of these claims.

Mortgage lenders require borrowers to carry it. The insurance is purchased with a one-time payment made on the front end of the loan, and is good until the loan is repaid. It protects you and your heirs indefinitely as long as you have an interest in the property. It will also cover loss and damages if the title is unmarketable (a legal term meaning having defect or likely to end up in litigation) or if there is not a right of access to the land. A title insurance policy protects you from:

- Errors or omissions in the deed
- Recording mistakes
- Forgery
- Undisclosed heirs
- Liens by contractors
- Liens for unpaid taxes

If there is a lien against you during the time that you own the property, your title policy does not cover you. You will be required to take steps to have the lien removed prior to selling the property. If you refinance the loan, you do not have to get a new owner's policy, but the lender will require a new lender policy. Even if you are refinancing with the same lender, the title insurance policy terminates when you pay off the mortgage.

Title insurance rates are not regulated in Alabama, the District of Columbia, Georgia, Hawaii, Illinois, Indiana, Massachusetts, Oklahoma, and West Virginia. In Texas and New Mexico, the state sets the price. In other states it might pay to shop for the policy.

Money Saving Tip: Ask for a Discount

If the property you are buying has been sold within the past five years, ask the title insurance company for a discount. You might also be eligible for a discount if you are a first-time buyer.

General Liability Insurance

As an owner-builder, the lender will require you to carry a minimum amount, $300,000 or $500,000, of general liability insurance for property damage and personal injury. It can be a comprehensive general policy or a broad form liability endorsement. Does this insurance seem like a waste of money? In 1970, Congress passed the Occupational Safety and Health Act (OSHA). The wording of this act in part reads, "...to assure so far as possible every working man and woman in the nation safe and healthful working conditions and to preserve our human resources." This relates directly to construction, where fatalities are commonly caused by falling from a roof or ladder, being struck by an object or by equipment, or receiving an electrical shock. It is vital that you carry liability insurance on your construction project.

Liability coverage protects the insured for damages that result from third party claims. These could be bodily injury, property damage, advertising injury, or personal injury. General liability covers both "premises operations" and the products completed, like construction defects.

It is important to go over the liability policy and understand exactly what it covers. Some policies exclude theft, wear and tear, machinery, testing, workmanship or materials, design error, collapse, flood damage, and earthquake damage.

Worker's Compensation Insurance

If you have employees, you will carry a workers' compensation policy. If you hire a contractor, he will carry the policy — unless all the labor and subs are independent contractors. If they are, they will fall under your liability policy if there is an accident. Lenders may require that you and the contractor, if you have one, sign a waiver stating that the lender is not liable for workers' compensation violations.

People who employ family members, or who hire a licensed contractor to oversee the project, do not have to carry workers' comp. You also do not have to have it if you use independent subcontractors. However, if you employ anyone else in any capacity, or if you participate in federal or state withholding for someone who works for you, then you could be required to carry workers' compensation insurance. Check with the IRS website at www.irs.gov in the section "Who is an employer?" to be sure.

Your subs, on the other hand, will have workers' compensation insurance for their own employees. They will also have builder's liability insurance, which they should show you. In addition to asking for their proof of insurance, you should also ask your subs for evidence of their license and bond.

Course of Construction Insurance

The course of construction insurance policy might also be called "builder's risk." Policies of this type vary, so it pays to check around. The policy should be in force before closing the construction loan or before materials arrive at the job site. The lender will be the payee in the policy.

This policy protects against theft, vandalism, fire, weather, and other kinds of damage during construction. It is sometimes referred to as an "all risk policy" because of its vast coverage. If some strange accident occurs, the policy will pay for the cost of rebuilding the home back to the way it was prior to the accident, up to the loan amount. The lender will require the policy.

Other Insurance

If your home has been designated as a Special Flood Hazard Area (SFHA), you must carry flood insurance. In addition, check with your insurance agent to see what kinds of policies are appropriate for your situation and your locale.

Sometimes people purchase "dwelling and fire" insurance instead of course of construction. After the construction is complete, the fire policy can be converted to homeowner's insurance at a savings.

Homeowner's Insurance

Once the home is complete, convert the fire policy or purchase a new homeowner's insurance policy. Be sure that you know the true value of the home, and that your policy will cover the total cost of replacement, should that ever be necessary.

What if your construction process damages the neighbor's property? Often during the course of construction, a neighbor sustains damage, such as a tree falling onto a fence or other property, or a sub or supplier driving onto the neighbor's lawn. There are many accident scenarios in which this type of damage can occur.

If a tree falls and it is not due to your negligence, the neighbors' own insurance policy will cover the damage. If you are negligent, though, your neighbor could file a lawsuit and your own insurance would come into play.

Ensuring Safety on the Job Site

Just as OSHA requires builders to create safety plans, so should you as an owner-builder. Safety should be the number one concern, not only to protect you and your investment, but also to safeguard the health and wellbeing of your workers.

You can keep "safety first" by operating just like a general contractor would. Safety meetings, referred to in the trade as "tailgate meetings," are informal, short discussions about keeping the work area safe from injury, fire, and other hazards. Utilize the following checklist to monitor the safety on your site:

JOB SITE SAFETY CHECKLIST:	
☐	Arrange for a portable toilet onsite.
☐	Provide fresh drinking water and disposable cups.
☐	Post Safety Rules onsite and also communicate them in writing or verbally to each trade contractor.
☐	Keep a first aid kit available.
☐	Post the phone numbers for police, ambulance, and fire station.
☐	Ensure workers and supervisors can get a reliable cell phone signal, especially in rural areas. If not, consider having a telephone line installed.
☐	Ground the temporary electrical service and all electrical tools.
☐	Check often to be sure electrical cords are kept away from water.
☐	Use only equipment that is listed, labeled, or certified. Use in accordance with manufacturer's instructions.
☐	Post "Warning" and "Danger" signs as appropriate.

☐	Insist on hard hats and steel-toed boots as appropriate.
☐	Cap protruding steel rebar, nails, and other protruding dangers.
☐	Ensure that power tools are well maintained and properly stored.
☐	Provide protective gear – goggles, gloves, and respirators.
☐	Require protective goggles when eye injuries are possible.
☐	Require wearing of personal protective equipment.
☐	Set a good example.
☐	Ensure there is adequate slope and proper fencing on edges of all ditches and trenches over 4 ft. deep.
☐	Place excavated material at least 2 ft. from edges of ditches and trenches.
☐	Cover open holes in sub-floors.
☐	Ensure safe access and use of all types of scaffolds.
☐	Install guardrails on all open-sided floors or platforms.
☐	Allow workers on the roof only with proper equipment.
☐	Construct a stair rail system on stairways of four or more risers.
☐	Remove excess and/or flammable scrap daily.
☐	Use only approved containers for storing flammable or combustible liquids.
☐	Insist that gas cans and other flammable liquids must remain in a secure area.
☐	Shut off welding tanks tightly when not in use – check and re-check.
☐	Monitor areas where soldering work is done – look for smoldering or burning wood.
☐	Insist that all workers maintain proper clearance from all power lines.
☐	Spread oily or paint rags outside to dry before disposal so they will not ignite.
☐	Keep a Material Safety Data Sheet (MSDS) onsite for hazardous chemicals.

Frequent, daily safety checks and clean-ups are most effective and will remind workers to be safety conscious.

Offer Supplies Onsite

Many experienced builders use a storage system of bins for items like nails, screws, and small tools that most of the subs will need. They feel that the amount of time saved more than makes up for the expense of filling and supplying the bins. Workers do not have to waste time running to the building supply store to get some item they need.

Owner-builders should incorporate this system into their building projects. Simply fasten several plastic storage containers together, or attach them to a framework of lumber. Locate them centrally on the site, and fill them with the basic equipment needed (there is a list at the end of this chapter). Make these available to the work crews when they come onsite. You will find that the site remains much cleaner, with fewer scattered items lying around.

Safety Inspections

Throughout the project you will encounter inspectors. The various inspections are covered in detail in Chapter 8; they are carried out by county and local officials, as well as inspectors who are sent by the lenders. These inspectors are not the enemy. They are there to see that your work meets structural standards and safety requirements. If the work fails inspection, you will be given time to correct it. You will then reschedule the inspection. You may be asked to pay an extra fee for the repeat inspection.

County and local officials normally stick to a set schedule of required inspections. For example, they may want to see the property before pouring the concrete footings, before pouring the concrete foundation, after framing, after the electrical/HVAC/plumbing rough-in, after laying the sewer line, and after the project is complete.

You can make several preparations to meet these inspectors with the best attitude. First, find out how much notice the inspectors will need. Find this out early, when you apply for the building permit.

Once you know when to notify the inspectors, always give them advance notice that you are ready for inspection. If you happen to make the appointment, and then find that for some reason you are not ready, be sure to give them a call so that they can reschedule.

On the day of the inspection, make sure the site is clean and organized. Be present during the inspection, so that if there is something you must correct you can learn exactly what it is and what steps you need to take to correct it.

Have a good attitude, even if you are asked to change something. These inspectors are only doing their job. You might even be able to learn from them.

Chapter 6 Checklist

Understand the different types of insurance needed and obtain:	
☐	Title insurance
☐	General liability insurance
☐	Workers' compensation insurance
☐	Course of construction insurance
☐	Flood insurance
☐	Homeowner's insurance
☐	Complete jobsite safety checklist.
☐	Conduct regular jobsite safety talks and inspections.

CHAPTER 7

Find and Negotiate with Subcontractors

A subcontractor (sub) is someone you hire to complete all or part of your project. Most of the subs you will work with are involved in specialty trades. For example, you might hire one person to build a block foundation, another to pour the concrete, and another to waterproof the basement.

The Importance of the Subcontractor

The sub's abilities and talents in his trade (or the lack of them) will make or break that particular aspect of your home. If a sub does a shoddy job, you will have to look at that every day that you live in the house. I remember a house I nearly made an offer on years ago. The wall in the bathroom had been poorly finished, resulting in an uneven finish and a crooked, thick corner instead of a sharp one. I asked the builder whether he could get the drywall contractor to come back and smooth the work; it should have been a one- or two-hour job to sand and respackle it. He instantly said no. Knowing what I do now, I suspect there had been more problems with the sub than just the bathroom walls. At the time, because of the short, sharp answer, I thought the builder was not cooperative; I chose to let the deal pass by. The builder lost a sale because of the sub's poor workmanship.

Money is at the forefront of everyone's mind when building a home, but choosing the right sub is much more complicated than just looking for the lowest price. Prices have a way of creeping up, though they never creep down! Most people focus too much on choosing the lowest bid, and that gets them into trouble. The right contractor will produce quality work, be reliable and show up when he says he will, and offer a reasonable price. He will not necessarily be the lowest bidder; in fact, he probably will not be.

Where to Find Potential Subcontractors

Locating a good sub is difficult, while finding a poor sub is easy. Those who do good work and are reliable are often busy, so be sure to start looking well ahead of the date you actually need them.

Local building supply

The easiest place to find the names of subs is at your local supplier. These people do business with everyone so they can readily tell you who is staying busy. Often there are bulletin boards where various workers can place their business cards. By visiting the supplier and talking with employees and contractors, you can pick up the names of several people who are staying busy at their trade. Remember, the subs who are busy, are typically the subs who are doing the best work.

Your carpenter

The lead carpenter you hire can provide a wealth of information about other subs. Carpenters are present on the job through more of the process than other subs, and as a result they know all the other professionals. Trust your carpenter to give you names of reliable workers.

Visiting job sites

Start looking for subs through your local building supplier or the carpenter you have hired. If you cannot find one this way, try visiting a few job sites. Some will have signs advertising the various subs; if not, simply find a house that is currently under construction, stop by, and ask questions. The subs on the site will most likely give you their contact information, references, and maybe even a price, all in a few minutes' time. If the boss is present on the job, consider asking him, too; far from being upset that you are there, he can offer some insight. Be sure to get his name and contact information so that if you do get in a bind or decide to hire someone to oversee this or a later project, you can get in touch with him.

The yellow pages

Traditionally, the yellow pages are not the best place to find subs. Many carpenters who work as independents are not listed. Other trades are also not listed as categories. The yellow pages do list most heating and air companies, plumbers, lumber suppliers, electricians, and some others.

Internet searches

The Internet has made it easy to find finding a listing of subs from your home or even on your phone. However, it can be difficult to know exactly what you are getting when you find a sub on the Internet. Search engines like Google and Angie's List can help you find subs in your area. Read user reviews and see what their customers have been saying about them. No matter how great the reviews sound, it is important to meet people and talk to them face to face before you involve them in the building of your future home. Remember, this is your home and, as the contractor, you are the boss. You should thoroughly interview and check the references of the people who will be constructing your house, just as you would for people who want to work for your company.

The Bidding Process

Get at least three or four bids on each job you want to subcontract. My policy is this:

- Every bid must be detailed in writing.
- Every sub must agree to sign a written contract.
- Every bid is for the job, not by the hour.

Step by step through the bid process

Keep names of potential subs in separate sections of your project notebook. When you have made all the design and materials decisions (as best you can, since some of this will require input from the sub), prepare a written set of specifications for the job.

Now, set up a file for each sub that you will be asking to bid. This file should contain a copy of your specs, a copy of your contract, and the bid sheet you want them to submit to you. Some of your subs will have their own computer software to create bids, or prefer to use their own forms. There is nothing wrong with this; just be sure that you have their workmen's compensation details and that there are signatures on the agreement you decide to use.

On the following page is a sample bid sheet. When you receive the bids, the most important item to look at is how detailed the specs are. If they are not detailed, you will need to go back to the contractor and ask questions to get a complete, accurate bid.

While you are creating files, set up a file of your own to keep track of bids. This one may be useful, although yours may have more or fewer categories:

SUB	NAME	BID	GOOD UNTIL	NAME	BID	GOOD UNTIL	BEST BID
EXCAVATION							
CONCRETE							
FRAMING							
ROOFING							
PLUMBING							
HVAC							
ELECTRICAL							
MASONRY							
SIDING							
INSULATION							
DRYWALL							
TRIM							
PAINT							
CARPENTRY							
FLOORING							
TILE							
GLASS							
GUTTERS							
ASPHALT							

Sample subcontractor bid sheet

Name:

Address:

Address of job that is up for bid (fill in your address):

Date:

Amount of Bid: ______________________________

Workman's Compensation Insurance Company Name:

Agent's Name:____

Certificate Number: ______________________Expiration Date:_______________

(Attach copy of insurance card.)

References:

1. ____

2. ____

3. ____

(List name, address, and telephone contact number for each reference.)

Work to be performed:

Materials you will supply:

Materials homeowner will supply:

Signature:____ Date: __________

Signature:____ Date: __________

Once you have gathered all the bids, discuss your plans, including your deadlines, with each potential sub. Make sure that by the end of the conversation you are clear about who will purchase materials, how the contract will be executed, and how payments are going to be made. If the written bid does not give a breakdown of exactly what materials are going to be used, find out.

Questions for subs

By asking questions of a sub, you both to get to know him (after all, he will be your employee) and gather information to make an educated decision. If you have never built or remodeled before, you may be at a loss as to what questions to ask. Besides the usual questions, ask these:

- Do you do the work, or do you send a crew? How big a crew would you send? How experienced are they?
- Why should I hire you over the other subs?
- What would you do if this were your own house?
- What are some ways I can save money?
- Is there anything I can do to save labor costs?
- What other contractors should I coordinate this job with? Is there someone you like or recommend to do that job?
- How far ahead should I schedule you?
- How long will it take you to do the job?
- What other suggestions do you have for me?

After you have talked to several subs, you might go back and ask more questions:

- Why is your bid so much higher (or lower) than everyone else's?
- What will I get for the extra cost?
- How could I save money?

One owner-builder cautions that you should not hurry after you receive bids. "People want to jump on the first bid or two, because they are impatient. They want to get the job done now. Plus, if you are doing an extensive remodel and you need, say, eight subs, and you want at least three to five bids for each, that could be forty bids."

That is fine. It does not matter that you have to create 40 sets of plans, or that you have to do 40 interviews in 30 days. Plan to spend the time necessary to get all the bids you need. Some builders allow the first month strictly for bid gathering, and the next month to revisit and negotiate each bid.

Email

Should you conduct part of your interviews by email? This can certainly cut down on the trouble and expense of creating plans, because you can fax from your computer or send them as email attachments. Plus, if you send out a request for bids with specific instructions in a fill-in-the-blank format, you can get lots of detailed bids quickly.

The disadvantage of using email for part of your interview process is that by not meeting face-to-face, you lose out on some of the meaning of the conversation. The advantage is that every discussion is recorded in writing; if a contractor does not follow through on what he promised, you have a written record of it.

How to Check Out your Contractors

We have all heard stories of contractors who grab the cash and disappear without doing the work. It is almost impossible to avoid this, especially if you build or remodel homes often. Here are a few suggestions that might help you prevent such problems.

- Visit construction sites or new homes and get the name of anyone whose work you like.

- Put yourself in the path of the trade. Join the Home Builders Association (HBA) and get to know the members; go to home and garden shows and talk to everyone in each booth about your project.

- Ask each sub for the names of other trade subs. In other words, ask the excavator whom he would recommend for the foundation; ask the framer who might be best for insulation.

- Check every name against the state registrar's list of contractors as well as the Better Business Bureau (BBB). It is common for contractors to have complaints on file, especially if they have been in business for a long time. One complaint is not a red flag for a builder with a large volume of work. But if the volume is small and there are several complaints on file, consider moving that name further down your list — or eliminate the name altogether.

Selecting the Right Sub

Now that you have written bids in front of you, and have talked individually with each sub, you should have a fairly clear assessment of them. Select the top three bids. This choice should be based equally on price and on your concept of the sub's character. Do not be afraid to eliminate someone if you have a bad feeling about him or her. One woman, who was her own contractor for a remodel on her home, relates the story of her shady sub:

> "I didn't want to use this guy, but the contractor who built the home insisted that he was okay. I had bad vibes all over the place. When he showed up to replace the windows, the dog – who never barks – started barking. And barking. The whole time the guy was there, the dog barked at him. I caught him cussing the dog and then kicking it. The reason I didn't put the dog up is because it made me sure I was right about the man. I only let him replace the one window and then I fired him. Just the one took him two entire days. Later I found that the window was not installed properly; it was crooked inside the frame, and wouldn't even open!"

After you have selected your final three, compare their bids against your original budget. It is amazing how variable bids are. You will find that some subs underbid the job. For excessively low bids, try to determine whether they are unable to estimate properly, they are new to the business, or they are frantic to get work. If one bid is more than 10 percent less than the others, and you cannot find a viable reason for it, eliminate that one. The same goes for extremely high bids; if the average bid is $6,500 and you get one for $9,500, that is too far away from the norm.

Is there a difference in the average of the top three bids and the price you expected to pay? If so, choose the sub you consider to be the top prospect and ask him what his best cash price is. Some subs will not change the price for a cash payment, but for others, money talks. Many subs expect to negotiate the price by 5 percent or more.

When you are sure you have negotiated the price as far as you can go without sacrificing quality, select the sub. Let him know right away that you will need a commitment for the given time period that you need him. You should have already outlined the working schedule so that you can tell him when that date is.

Contractual Language

Terms and Conditions: All agreements and oral promises between you and each contractor should be put in writing. This is done to protect both the customer and the contractor. If you intend to do some of the work yourself, or if you are hiring more than one contractor for jobs that would normally be done by the same person, those details need to be entered in the contract as well. At the least, the written contract should include:

- A thorough description of the work to be done
- A complete list of all materials to be used — quality, quantity, weight, color, size, brand name, and so on. Use part numbers whenever possible.
- The total cost, including a breakdown of charges for both labor and materials
- The start and completion dates that you have agreed on
- A payment schedule

- A right-of-rescission clause, which states your right to cancel the contract within three business days. This clause should also state that the contract is null and void, or can be renegotiated, if the job uncovers unexpected or hidden problems or damage after work has begun.

- Warranty clause: Any warranty offered on products or services by the contractor should be in writing. Study the warranty carefully; make certain you understand all the terms and conditions, including the length of the warranty. The warranty must state whether it is a FULL warranty that gives the consumer certain automatic rights, or a LIMITED warranty that restricts certain consumer rights.

In addition, make sure the written contract includes:

- The contractor's full name
- Their address
- All telephone numbers (home, work, cell)
- Professional license number

Also, make sure that the contract discusses debris and material removal once the job is finished. Make sure you both understand who is going to remove it and who carries the cost.

Once you have a completed contract, go over it with each sub, making changes together as needed. Never sign a partial or blank contract. Read every contract clause carefully and ask any questions you may have before signing. Both parties must initial any changes to the typed agreement. Get signatures and make copies; retain one for your office and an extra for yourself, and give a copy to the sub.

Be sure to stay in contact with your subs, especially if it is going to be some time before they are needed. If the schedule happens to change (and it will), let each sub who is affected by the change know as soon as possible. This is the only house you are building, but it is not the only one your sub is working on. By knowing your plans, he can adjust his schedule and still get paid — and he is much more likely to be available when you are ready for him.

Paying Subcontractors

When the written contract is hammered out between you and the sub, it will contain a price for the work. This may be a certain number of dollars per sq. ft. or a lump sum. While you are writing out the contract, work out a payment agreement. Do not be shy about asking the sub how he is paid. Subs discuss this with every person who hires them.

Never agree to pay the full price up front. You will be left with no leverage if you do this. Instead, the written agreement should normally call for draws. This is expected, and the best way to ensure that a sub will finish a job is to withhold money. Do not hold back money that you owe the sub, but do not pay everything up front. By agreeing to certain percentages as you go, you can be sure that the job gets completed.

Here is an example that will show you why this is important. Imagine that you have hired someone to put in your heating system. Halfway through the job, he goes broke. If you have paid him 75 or 100 percent of his total bid, you will not have enough left over to pay a new HVAC installer to finish the job. You are left without heating and cooling, without the funds to pay for it, and perhaps with an inspector is coming out tomorrow. What will you do?

Here is another example. One owner-builder hired an electrician to do all the wiring on his home. The man asked for 60 percent up front and 40 percent at completion, which seemed reasonable — until the county inspector showed up. Some of the wiring did not meet code, but Don had already paid the electrician; his phone calls were not returned. Don paid an extra $2,000 to a different electrician to resolve the issues.

For these reasons, there needs to be some sort of arrangement so that you can safeguard yourself, and at the same time pay your sub for his work. Below is a rough guide to the draws you might agree on. Remember that these may vary; different subs may have established different plans that works for their companies, and trends may vary by geographical location as well.

Subcontractors' draws

It is very important to decide up front with the sub how he will be paid — and then stick to it. Do not pay a sub for work he has not yet completed. Some subs live for their Friday afternoon beer money, and others are hard-working family men who would not dream of skipping out on the job before it is finished. You have no way of knowing which one is the sub you have hired, unless he is your brother. Assume that he is the beer money fellow, and do not pay him if you have not agreed to do so. Of course, if you owe a certain percentage according to your agreement, go ahead and make the payment on time, without waiting for him to ask for it.

Do not pay for work that has been left incomplete when the sub has told you it is complete. Always inspect to be sure this is not the case. Sometimes people in a hurry to receive the paycheck rush through the job. Make it a policy to always inspect the work before paying; as the subs see you doing this they will pass the word that you will not pay unless the work is done.

Do not pay the entire sum up front. If you pay your sub he will never come back. Agree on a certain percentage up front if he requires it, then fill out a schedule of the work he will perform and the way he will be paid.

Carpenter's draws

The carpenter's draw is a little different from everyone else's. A carpenter is paid as he does the work. This will take some communication on your part, if you do not have the payment agreement written in a specific way in the contract. The carpenter gets a draw based on the percentage of work he has completed. So if he is about 30 percent of the way done, pay him a 30 percent draw.

Plumbers' and electricians' draws

Plumbers and electricians are paid 40 percent upon completion AND INSPECTION of rough-in, the remaining 60 percent after the job is complete and you have had the final inspection.

Brick masons' and painters' draws

These subs will require a draw during the process. It is best to inspect the job and compare what they have done so far to what remains to be done. Do not pay ahead; instead, pay in smaller increments if needed so that you stay right in line with what they are doing.

Getting Results from Your Subcontractors

One of the fastest ways to get a sub to complete the job — and maybe even to do a great job — is by telling him that you will pay cash. Cash is a good incentive to get things done. If you are concerned about the IRS, write out a check and then cash it, recording it as his pay so that you will have a paper trail showing that you had the work done.

When you think about getting results from your subs, remember that the results you are looking for are only what you have gotten in writing. If you are looking for specs that are not in the contract, you are not likely to get them; technically you did not ask for them. Be sure that the wording in the contract matches your expectations.

Paperwork

Many subs hate doing paperwork. You may find that it is difficult to get some to write out bids or to sign documents. Do not take this personally; it is merely a fact of working with certain trade professionals. The only documents that are necessary are the contract and the invoice marked "paid in full." If you absolutely must hire someone who works this way, consider writing out the contract yourself, and insist that he present you with an invoice marked "paid." These two steps will protect you should problems arise later.

Contractor Checklist

This is a list of some of the potential contractors that will be needed in the course of construction:

☐	Architect
☐	Key Carpenter
☐	Surveyor
☐	Grading and excavation contractor
☐	Soil treatment company
☐	Footing contractor
☐	Brick mason (to build block foundation)
☐	Concrete (walls, slab/floors, driveway, walkway)
☐	Waterproofing (basement)
☐	Electrical
☐	Plumbing
☐	Well and septic
☐	HVAC
☐	Roofing contractor
☐	Insulation contractor
☐	Drywall contractor
☐	Paint contractor
☐	Finish carpenter (You may use your main carpenter for this.)
☐	Flooring contractor
☐	Countertop contractor
☐	Landscape firm
☐	Fee appraiser
☐	Tile contractor
☐	Cleaning firm

Chapter 7 Checklist

☐	Generate a list of potential subcontractors for each craft (See list above.).
☐	Get referrals and recommendations.
☐	Prepare bid requests and get bids.
☐	Check out leading bidders.
☐	Select subcontractors.
☐	Write contracts including payment schedules.
☐	Get all necessary signatures and make extra copies. Store copies separately from each other, or scan into computer.

CHAPTER 8

Renovating a Home

There are many good reasons to renovate a home. If you have decided that a renovation is the best choice for you, proceed in the same way that you would to build a new home. Create a budget, determine what needs to be done, and interview contractors.

The Home Appraisal

When renovating or remodeling, take a couple of extra steps to protect your investment. The first step is to have the home appraised. Explain to the appraiser that you are planning to renovate, and let him or her give you the value of the home when the remodeling is complete.

The Home Inspection

Another important step is to have the home inspected by a professional, who can tell you exactly what condition the home is in before you begin. A comprehensive inspection will give the status of critical issues in your home. For example, is the electrical circuitry safe and polarized correctly? Is moisture getting into the house? Are structural supports in place as they should be? What is the condition of the firebrick on the chimney? Is the chimney properly aligned? Finding out about these things may change the scope of your renovation, and will alert you about hidden problems before you begin to remodel. Then you can plan for them in your strategy and your budget. When people dive into a project without an inspection, they often uncover problems as they go and have to react to them, rather than being proactive. The few hundred dollars you spend on the inspection will be quickly repaid by the time you save on the project.

Note that a home inspection is not like an appraisal. It does not tell you what the house is worth. Instead, the inspection tells you about the condition of the home. It tells you whether there are components of your home that need major repair work, or must be replaced.

At the American Society of Home Inspectors' website, you can find a "virtual home inspection" link which will walk you through a step-by-step inspection. This will show you why inspections are useful. Are the walks safe? Is the plumbing keeping water and gasses inside the pipes and away from the interior of your home? Often problems with pipes cannot be seen, but rust in a pipe will accumulate and eventually restrict the water flow.

- **Electrical** — Are breakers, fuses, and disconnects as they should be? Is the service enough for the house?

- **Heating** — Are the vents, flues, and central and through-wall air conditioning safe? Make sure there is an overflow plan so that condensation is carried away from the house; on the interior, study the safety of both the insulation and the ventilation.

These are just a few of the many details an inspector will look at. You should plan on the inspection taking as long as two or three hours.

Armed with the inspection and the appraisal, make a decision about the project. Remember, if the appraised value after the renovation will make your home the highest priced house on the block, it will be very hard to sell in the future. You could even lose money when you try to sell it, because you might end up accepting a price that is more in line with the prices of the neighbors' houses. Remember that your renovation costs may end up being much higher than you anticipate, so the value comparison may not be realistic.

Environmental Testing

If you did not have tests done on your home when you originally purchased it, it may be beneficial to test for the presence of hazardous materials that could jeopardize your family's health. The State Department of Environmental Protection is a good source for learning about testing in your area.

Radon is a radioactive gas that occurs naturally in soil. It is the second leading cause of lung cancer in the US, and nearly 1 of every 15 homes has elevated radon levels. It does not matter whether your home is new, old, sealed, or has a basement; the gas rises from the ground to the air and into your home through the foundation, through cracks, gaps, and construction joints. If you or your inspector find high levels of radon in the home, it is important to seal foundation cracks and other openings. Installing a vent pipe system is also a common way to remove the radon from under the home and vent it to the outside. If your renovation involves the conversion of an unfinished basement to a living area, test it for radon before you begin; the added cost of repair, if you find radon, is only $1,000 to $2,000. Radon can also enter through well water. Sometimes building materials give off radon, although this is rare.

Asbestos, which was once used to insulate homes as well as for flooring, is present in nearly every house that was built between 1920 and 1960. Asbestos can be found in siding, flooring tiles, and pipe insulation. It does not have to be removed. If it is left alone and in good condition, it does not pose a health hazard. But if asbestos is damaged by your renovation efforts, it must be abated or removed. Asbestos can be covered over with new material if you are careful not to break it, and if it is completely covered by the new material. If you must remove and dispose of your asbestos, it is expensive. A licensed professional must do the removal.

Another substance found in pre-1960 homes is lead. You can find lead in two places: the paint and the water system, where lead solder was used on the pipe joints. Lead poisoning is dangerous, especially to young children. In spite of the best efforts of realtors and the government, there are still children becoming ill due to lead-based paint. One child, who was found to have 10 times the allowable amount of lead in his system, was determined to have gotten the paint from a window casement where the family often opened and closed the window. He was not putting the paint in his mouth, but because his face was at the level of the window ledge and the paint was peeling, he still became contaminated by the paint.

You can perform a lead paint test yourself with a kit. You can also hire a licensed professional, who is able to recommend ways to remove the lead if it is found. Lead-based paint can be easily covered over with special paint.

If there is lead of over 15 ppb, or other contaminants in the water, contact your local health department. They will have information on professional remedies. These are steps you can take to prevent poisoning:

- Let the tap water run for 15 to 30 seconds before using it.
- Do not cook with the water from the hot tap or drink it; hot water dissolves lead faster than cold water.
- If you are building a home, remove the faucet strainers from all the taps and run the water for 3 to 5 minutes.

If the copper pipes are joined by lead solder and it has been done since 1986, then the plumber who did the work did it illegally. You can ask that he replace the lead solder or you can notify the health department.

Determine whether the service line connecting your home to the water line is lead. You do this by asking the home inspector or the local building inspector to look at it. However, municipalities are not required to remove lead water lines.

During renovation projects, lead dust can be inhaled. Lead can also enter through drinking contaminated water, or when small children eat paint chips.

Other Possible Hazards

Numerous other causes can lead to environmental hazards or "sick" houses. If you are renovating the home you already own, or you have purchased one for that purpose, pay the small fee for a professional inspection. Here are some of the possible dangers:

- Aluminum wiring
- Chimney and flue defects
- Drinking water – arsenic
- Asbestos
- Fiberglass
- Past flooding
- Electromagnetic fields
- Oil storage tanks, either above or below ground
- Insulation or ventilation issues
- Sewage or septic backups leaks and clogs
- Structural defects

Extreme Makeovers

If you are considering a major overhaul, there are pros and cons. Remodeling means a lot of hard work, subs that you do not know in and out of your house at all hours, and heavy expenses. Will you recoup your investment? Maybe. In most cases, you can expect a return of 60 to 80 percent on your investment, depending on what job you undertake.

If you sell the home instead, there are still expenses: real-estate fees, moving costs, and a few thousand dollars in closing costs. So when does it make sense to remodel?

Return on Investment for Some Remodeling Projects

Remodeling Magazine calculates the cost versus value of remodeling projects every year. Although the actual price varies depending on which area of the country you live in, you can get an idea of which projects might be valuable for you. Here are some of the more popular renovations:

MID-RANGE PRICE	COST RECOUPED
Attic Bedroom Remodel	76.6 percent
Deck Addition	85 percent
Family Room Addition	68 percent
Siding Replacement	83 percent
Window Replacement	80 percent

For more information, visit REMODELING online at **www.remodelingmagazine.com.**

When Feelings Interfere

The emotional investment when remodeling a home that you already own cannot be measured. If you live on the same street as many of your relatives or closest friends, for example, you may not care about returns on investment; you simply want your home a certain way so that you can live there more happily.

Sometimes feelings go the other way — you are sick of living in your house, and you want to move on. If this is the case, consider building your dream home from scratch. Remodel the home you live in only enough to make it as marketable as you can. Use the rate of return guide above) to decide which projects you will undertake. If the return brings the value of the house up, it is a good investment. In some cases the renovation brings a dilapidated house up to the level of the neighborhood. If you are in this situation, the renovation is almost a must, unless you are willing to accept a very low price for your property.

Sometimes you love your location, just not your house. There may be traffic flow problems, water problems, or mold. It might simply not fit your ideal. Gutting or tearing down a house may not always be appropriate, but there are times when the location of the house or the beauty of the lot supersedes the desire to buy something else. If you are thinking of tearing down the house and starting over, there are a few things to think about first:

- Do you have a mortgage on the property? If so, it is illegal to tear down the house. One lender said that his institution would allow it, but only if the total balance owed on the loan is less than the value of the land — because the land will be all that is left.

- How restrictive are the zoning laws in your area? Do you have limitations in your Homeowners' Association covenants? These will limit your remodeling options, and may cause you to rethink doing a major remodel. If you decide to do a major renovation, remember that many municipalities charge a lower property tax rate on remodels as opposed to new construction. Leaving part of the old home standing can save money at tax time.

- Besides the restrictions imposed by community covenants, there are the neighbors to consider. In most cases, the planning department requires that the community be notified of major renovations. The neighbors then have a voice in the decision. Sometimes there is even a design review board that must approve your plans. Be sure that your neighborhood will tolerate your plans; whether you are going to underbuild, overbuild, or fit perfectly into the neighborhood design, their tolerance matters.

- How long do you plan to be in the home? If you are going to be there for a long time, you are improving your lifestyle by doing a major remodel. If you are not staying in the home for a long time, or you are unsure how long you will be there, it will be more important to assess the impact of the renovations against the resale value.

- Can you find a reasonably priced house with all the amenities you want? If the supply of homes in your area is limited, and you already enjoy your home, a remodel is the perfect choice. You can make the house exactly what you want it to be.

- How does the house compare to others in the neighborhood? If it is in need of siding, has only one bathroom where all the neighbors have three, needs updating, and has no distinctive features that will make it stand out from the crowd, then it needs a remodel, whether you plan to stay there or not.

If you do decide to demolish an existing home, do not forget to add the demolition and removal costs to the expense of your project. If the property contains hazardous waste, which is common in old houses, it can cost even more. Get an estimate for both the demolition and the removal of all materials before making a firm decision.

Remodeling Checklist

☐	**1. ANALYZE YOUR BUDGET.**
☐	**2. ANALYZE THE COSTS.** Be sure that you include: • Permit fees • Construction materials • Labor costs (you need to know the total hours) • Tools and equipment • Safety equipment • Materials • Incidental repairs, which are inevitable when you delve into an existing structure • Cleanup, including rental bins, trash removal or dumping. • Decorative items • Other expenses
☐	**3. CONSIDER THE RETURN ON INVESTMENT.** • Market valuation – If the home prices in your area are increasing, the renovation may add a good value to your home. If the market values are decreasing, the renovation might not add value to your home. • Home comparison – How will the renovation affect the "look" of your home as compared to others in the neighborhood? If your project does not fit, in some instances the value is decreased.

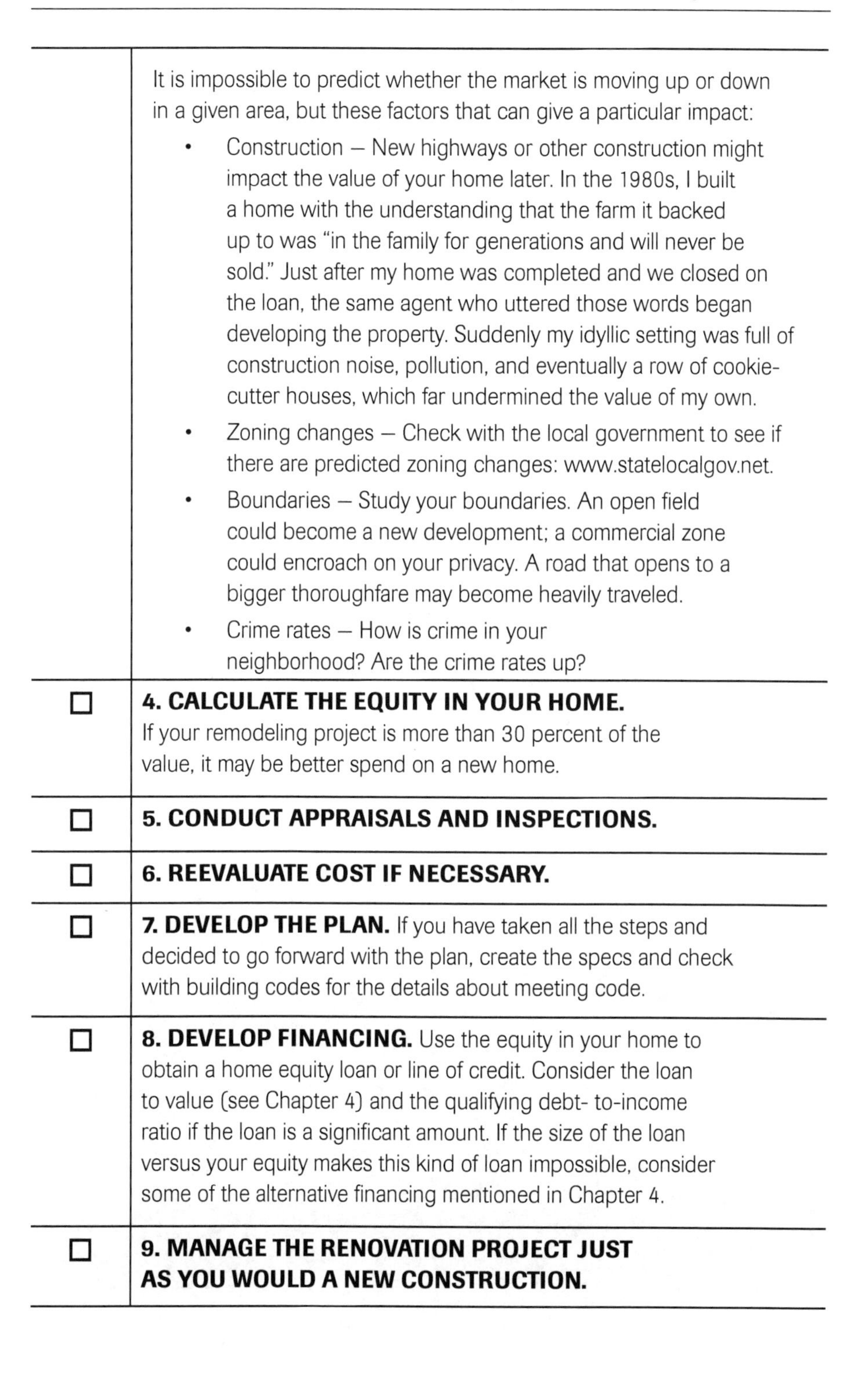

	It is impossible to predict whether the market is moving up or down in a given area, but these factors that can give a particular impact: • Construction – New highways or other construction might impact the value of your home later. In the 1980s, I built a home with the understanding that the farm it backed up to was "in the family for generations and will never be sold." Just after my home was completed and we closed on the loan, the same agent who uttered those words began developing the property. Suddenly my idyllic setting was full of construction noise, pollution, and eventually a row of cookie-cutter houses, which far undermined the value of my own. • Zoning changes – Check with the local government to see if there are predicted zoning changes: www.statelocalgov.net. • Boundaries – Study your boundaries. An open field could become a new development; a commercial zone could encroach on your privacy. A road that opens to a bigger thoroughfare may become heavily traveled. • Crime rates – How is crime in your neighborhood? Are the crime rates up?
☐	**4. CALCULATE THE EQUITY IN YOUR HOME.** If your remodeling project is more than 30 percent of the value, it may be better spend on a new home.
☐	**5. CONDUCT APPRAISALS AND INSPECTIONS.**
☐	**6. REEVALUATE COST IF NECESSARY.**
☐	**7. DEVELOP THE PLAN.** If you have taken all the steps and decided to go forward with the plan, create the specs and check with building codes for the details about meeting code.
☐	**8. DEVELOP FINANCING.** Use the equity in your home to obtain a home equity loan or line of credit. Consider the loan to value (see Chapter 4) and the qualifying debt- to-income ratio if the loan is a significant amount. If the size of the loan versus your equity makes this kind of loan impossible, consider some of the alternative financing mentioned in Chapter 4.
☐	**9. MANAGE THE RENOVATION PROJECT JUST AS YOU WOULD A NEW CONSTRUCTION.**

Chapter 8 Checklist

☐	Have the home appraised and inspected.
☐	Have the home environmentally tested.
☐	Review the "Other Possible Hazards" list and determine which need to be checked.
☐	Review your potential investment vs. resale payback vs. other alternatives.
☐	Develop the plan.
☐	Develop the budget (see Remodeling Checklist above).
☐	Finance the renovation.
☐	Manage the renovation as you would new construction.

CHAPTER 9

Comply with Building Code, Inspection, and Permit Requirements

Building codes exist in every jurisdiction in the US. Some are stricter than others; some have the same rules but interpret them differently. Most building departments rely on a Uniform Building Code (UBC), at least to some extent. Some must create special rules because of special conditions that are indigenous to your area, like snow loads, earthquakes, and hurricanes. This can make it hard for you, as a new builder, to interpret and follow the building codes. Get to know the people who will be handling your project at the local code enforcement office.

Design Review: Understanding What is Allowed

If you are building in a neighborhood with many homes that all look alike, you can assume that the city, county, or neighborhood commission has developed specific guidelines governing what can be built there. If you plan to build in such a community, expect strict guidelines and much delay. Begin by getting a copy of the local building code from the local government, and the formal guidelines, called CC&Rs, for the subdivision.

Before spending thousands of dollars for formal plans, get the preliminary design approved. Make sure that everything you have in mind will pass the approval, and change the design to fit if you find that it will not pass. This will save a lot of headaches later on — and possibly save you from having to tear down part of your house.

Building Permits

Homeowners are able to get all the permits needed in most jurisdictions in the US. It is not necessary for you to get all of them; many of the licensed tradesmen you hire as subs will provide their own permits.

To get a building permit, you will supply two sets of prints and specs to the local code enforcement office. The authorities will examine the plans and either approve or reject your project.

After the plans are approved, one set of prints will be given back to you. The other will be kept in the code enforcement office. You must keep the approved set of plans at the job site at all times. It is stamped and marked as approved; the inspector will need access to it when he comes onsite.

In addition to the plan approval, the code enforcement office will issue you a permit. This has to be displayed prominently at the job site. Wrap it in plastic to protect it, and mount it high on a tree or pole so that it remains safe and unscathed by knocks, bumps, or vandalism.

Some of the other permits may be kept in your files; others must be posted. The plumbing, heating and electrical permits, which will be acquired by your subs, must not only be posted, but be visible from the road.

Permit Checklist

The required permits vary from one area to another. Permits that will be necessary for your project are:

- Building permit
- Plumbing permit
- Heating permit
- Electrical permit
- Septic permit
- Driveway permit

Inspections

Inspections were covered in Chapter 6, but they are so important for many owner-builders that they are mentioned again here. Getting plans approved and proper inspections done can go smoothly, but it often becomes a nightmare. Unless your regular job is in the realm of architecture, engineering, or contracting, you do not know the process or the people who are involved.

Learn all that you can. Familiarize yourself with the rules and regulations of your local government, and follow the required steps, called a "plan check." Be prepared to provide all your plans and whatever else they ask for. It is helpful if the contractors you are using are familiar with these departments, and know to whom you should submit documents and how to streamline the process.

One important inspection is the one that precedes the driveway permit. If this is required in your jurisdiction, an inspector will look at the spot you have chosen for your driveway to enter the main road, and determine whether the location is safe.

If it is, he will tell you what size culvert pipe you need and issue you a permit. If the location is not the best due to the lay of the road, and you may be asked to move the driveway.

Failure to have any of the necessary inspections performed can result in extra work and expense. For example, if you were asked to move the driveway, imagine what it would cost to remove the concrete and replace it. If you forget your plumbing inspection, you could be asked to remove walls and fixtures to allow the inspection to take place.

Building inspectors

County inspectors will review the work done and certify it at certain stages of the remodeling or construction. For example, temporary electrical service will be inspected to ensure that it is properly grounded. The slab will be inspected prior to pouring concrete to be sure that it is insulated correctly. Footings will be inspected to be sure that they are resting on solid (load-bearing) ground. It is important for you to be present at every scheduled inspection. The inspectors from the county or municipality are not inspecting the quality of the work; they are making sure that it meets code. The final systems like insulation, electrical, plumbing, HVAC, and others must be checked to ensure that they:

- Work
- Meet code
- Are safe

What kind of codes might be enforced? Some areas restrict the size of homes to one or two stories; some require that driveways be asphalt or cement. Some restrict lots to a certain size or kind of house. Homeowners' Associations might state that you have to use stucco, or not use stucco. They might require a certain size of house or a certain size window. These requirements are all found in the CC&Rs that are a part of the land.

After these final inspections take place, a certificate of occupancy will be issued.

Bank inspectors

The inspectors who represent the lender(s) arrive at the end of each phase; their purpose is to see that the money that was borrowed is being used for its intended purpose. These are not building inspectors, so they are not looking into quality or code. They are only taking photos and keeping records.

Most subs schedule their own separate inspections; you should also attend these inspections. How do you make sure that the work done by subs passes inspection, meets code, and is also exactly what you requested? The simplest way is to hire your own, independent inspector. This person can pass the work before you make any payments to the sub. You are completely within your right to withhold payment until the sub has delivered all the services that are listed in your contract. By hiring a professional inspector, you both protect your investment and create a paper trail, should the need for one arise. Inspections highlight the importance of a detailed, comprehensive construction contract, and the need to thoroughly discuss all project plans with subs.

When construction fails inspection

If any step of your construction fails its municipal inspection, it means that step did not meet code. You simply make the corrections or adjustments and request that a follow up inspection be conducted. There will be an additional re-inspection fee. If your independent inspector finds fault with any part of the construction, he must explain the reasons, which you can use to discuss the need for corrections with your sub. Independent inspectors do not fail items, they simply point out the physical conditions and show you what parts may need replacing or repair.

Remember that inspections have a positive, not a negative purpose. There might be as many as 15 inspections during the building process. They may seem difficult to schedule, and you might often feel as if the inspection process holds up the project. But inspections ensure that the work is done properly, and they also protect your investment.

Chapter 9 Checklist

☐	Get copies of local building codes and study these in respect to your plans.
☐	Review the Permit Checklist to insure you have all needed permits.
☐	Understand what inspections will need to be made and what the inspector will be looking for. Remember, they are your friends!

CHAPTER 10

Building Materials and Foundations

Where to purchase building materials

To purchase materials, you first need a list of all the supplies you might need. This is called a "take-off list." This chapter includes a master list to help get you started. Supplier prices, as you are about to discover through personal experience, vary wildly. You will need to get several quotes on each kind of supply to find the best prices.

Where will you get the quotes? If you have already selected subs to do the work, ask them where you should purchase your materials. They often know who has the best pricing or who will offer a discount. Do not be tempted to let the sub buy the materials for you, in case there is a markup. Instead, go to the supply house and negotiate your own price.

Often when you purchase plans, a materials list is supplied for an extra fee. You can use this list as a guideline, but it is not infallible. Building codes and geographical climates create differences in the materials used in various parts of the country. Even the wall studs vary; some areas may use 2X4s and others use 2X6s — so if your list calls for one instead of the other, you will end up with the wrong supplies.

Should you let the supplier create your list? As a test, try this. Choose several suppliers and ask each of them to create the take-off based on your plans. Most suppliers are willing to do this for free to entice you to do business with them. In the meantime, create your own list of supplies based on the list in this book. When you gather all the lists, sit down and compare them. You will find variations; in some cases, they will have forgotten to list certain items. In others, the prices or amounts figured will vary. There are several ways to do a job, but it is better to create the list yourself.

Building materials may come from:

- Lumberyard
- Builders supply store
- Do It Yourself superstore
- Builder's closeout store
- Item-specific warehouse
- The Internet
- eBay

How to Save on Building Materials

Shopping for building materials is just like shopping for any other consumer product. You can buy in bulk and save; you can buy at the superstore and save some; or you can buy at a specialty store and pay premium prices. The prices you pay also depend on your purpose. For example, a professional builder considers time a more important factor than price, so as long as the price he gets is fairly competitive, he is not going to spend a time shopping around for a special deal. He could make a little more profit by comparison shopping, but if he is earning, say $30,000 profit building your $275,000 house, then he would prefer to move forward by building rather than shopping. Plus, the builder may be earning as much as 30 percent on the items you select for your home.

As an owner-builder, you do have time to look for bargains. You can shop and save — the more time you are willing to invest in the search, the more you will save. Some contractors you speak to will try to tell you that they get a "builder's discount" that you cannot match, but this is not true. Individuals can find terrific savings on any number of materials by simply looking for deals. Some of the savings owner-builders have mentioned include:

- **Staircase parts:** $14.99 at the local builders supply, $12.99 at Home Depot, $$3.75 at the builder's surplus store 20 miles away. The homeowner needed 98 of these; the savings totaled $1,101.

- **Tile:** One homeowner located a tile supplier four hours away. Since they were installing 1,800 square feet of tile flooring, he and his wife decided to make the drive. By choosing closeout tile, they paid $0.99 per square foot instead of $11.88 — for a total savings of $19,602! This was well worth the eight-hour round trip drive to the supply store.

- **Windows:** High quality windows are costly, yet it is possible to find terrific deals. I ran across a couple whose renter had ordered $36,000 worth of custom windows — then skipped town, leaving the owners to pay the bill. They sold the windows for $3,600 just to get them out of the way. A self-employed installer altered the design of the 15 windows in the house and put them in for $1,800. An added bonus for the homeowner was that switching to bigger windows let in a lot more sunlight.

- **Appliances:** Do not put off shopping for appliances until the job is nearly complete. It is possible to find closeouts, end-of-year models, and other deals. My own latest washing machine was purchased at Sears for $300 less than the listed price. Its model number was the same but the SKU number differed from the one that was on "sale" for $150 off; however, there was no difference in the product, only in the one number, which was printed in tiny writing on the display at the store. They had two of these still in the warehouse; I presume it was simply last year's version of the same washer. No employee could explain why it was different. I had to ask them specifically to type in that special SKU number to get the right price. After I asked, they even threw in free delivery. Keep your eyes open and check those part numbers!

- **Siding:** Siding is one of the products that has the biggest variation in price (windows are another). Siding quotes vary as much as 30 percent from one supplier to another. Subs are especially helpful in telling you where to shop for siding. One homeowner purchased through an individual who sold siding on the side and saved 50 percent of the previous lowest quote he had received.

There are savings on every single item in your home; you just have to be willing to shop and use your head. Watch the newspapers for sales, check out all the local closeout stores, and do not be afraid to talk about your project and ask questions. One homeowner talked with a builder's supply store manager, who did not offer the best prices. But the manager remembered him, so when some windows were returned to the store, he phoned the homeowner — who got them at 25 percent off the retail price.

If there is an area fairly close by where wages are lower, shop there. Even chain stores have different pricing for different geographical locations. Driving an hour or two to save 10 cents per part does not make sense, but when you are purchasing flooring, pipes, windows, or sheetrock for an entire house, you could be saving thousands of dollars.

Should you purchase all the materials from one supplier?

This is a question to seriously consider. It is not the best way to save money. Often you find that one supplier has great prices on lumber, but not on siding. Or it has good prices on doors and windows, but the highest quote for shingles and trusses. However, there are several advantages to using just one supplier. You will get better service if they are supplying the entire job. You can ask for a price lock, so that the price quotes you get up front are good for the duration of the project. And, with only one supplier, the paperwork is easier; you only have one set of invoices, and one place to make returns or discuss shipment errors.

Whether you decide to use one supplier or several, always ask questions. Ask whether that is the best price; ask whether you can get an extra 10 percent off if you purchase all the parts at this location. Ask for free delivery. Ask to see the discontinued items, damaged stock, and overstock.

It is always possible to save by purchasing the materials separate from the labor. If subs are giving you bids with the materials included, they are using a formula to compute the cost of materials. By looking at the breakdown and comparison shopping, you can beat that price — then find a sub who is willing to do labor only. This saves money on the product you are purchasing, and allows you to control the quality of the materials. You will know exactly what you are getting. You will have the warranty information at hand, and you will be able to get a refund if you have overestimated the amount of material that you need.

You might even save on labor; you will see that you can install some of the components by yourself, and you can negotiate with subs for labor-only until you find a good price.

Sample Take-off List

TAKE-OFF LIST INFORMATION		
Plan Name / Date	Sample	
Plan Description	Residence	
Customer / Phone	Private Customer	11.222.35
Contractor / Phone	D&D	800-555-1212
Supplier / Phone	Sample Supply Store	111-555-3333
Deliver To	8766 Pine Tree Ln, Mytown, Mystate	

FOUNDATION / DRAIN MATERIALS				
21	pcs.	AB0812	Anchor bolts	for hold down
77	pcs.	1/2 x 8"	Regular J-bolts	mud sills
18	pcs.	16 x 8 x 8"	Vents	crawlspace
9	sheets	2" R10.5	Rigid board 4x8	slab insulation
250	lin. ft.	4"	ADS solid pipe	downspout drain
235	lin. ft.	4"	ADS perforated pipe	perimeter drain
3	pcs.	4"	Connectors	drain pipe
8	pcs.	4"	Elbows	drain pipe
16	pcs.	4"	Tees	drain pipe
4	pcs.	4"	Ys	drain pipe
SILL PLATES				
236	lin. ft.	2 x 8"	Treated sill plate (check code; 2X8 gives more stability than 2X6)	Mud sills
5	rolls	6" x 50'	Sill sealer	Mud sills
132	pcs.	1/2"	Nuts & washers	for anchor bolts
BASEMENT FLOORS				
1	pcs.	9'	Steel I-beam	Basement
1	pcs.	8'	Steel post (if needed)	Basement
WALLS - BASEMENT				
172	lin. ft.	2 x 6"	SPF	plates

102	pcs.	2 x 6"	SPF, 92-5/8"	studs
64	lin. ft.	2 x 6"	SPF	blocking, bracing, etc.
86	lin. ft.	2 x 4"	SPF	plates
32	pcs.	2 x 4"	SPF, 92-5/8"	studs
44	lin. ft.	2 x 4"	SPF	blocking, bracing, etc.
20	sheets	3/4"	Birch Plywood, 4x8	sheet siding
9	rolls	1000 sq. ft.	Typar	building paper
2	roll	2"	Tape	for building paper
54	tubes	10 oz.	Silicone Latex white	caulking
4	pcs.	HDA		Holdown
16	pcs.	5/8" x 4-1/2"	Stud (hex) bolt set	for HDA
4	pcs.	5/8" x 24"	Steel threaded rods	for HDA
8	pcs.	5/8"	Nuts & washers	for HDA
18	pcs.	HD5A	Holdown	
40	pcs.	3/4" x 4-1/2"	Stud (hex) bolt set	for HD5A
20	pcs.	5/8"	Nuts & washers	for HD5A
20	pcs.	5/8"	Nuts	for threaded rod
14	pcs.	MST	Metal strap tie	for walls
1	roll	CS16	Metal strapping	for walls
13000	pcs.	16d	V.C. Sinker, Hitachi	Framing nails
40	lbs.	1-1/2"	Teko nails, galvanized	Connector nails
25000	pcs.	8d	Galvanized nails	Sheathing nails
FLOORS - 1ST LEVEL				
2180	lin. ft.	11-7/8"	TJI	joists
386	lin. ft.	1-1/4" x 11-7/8"	Timberstrand	rims
118	lin. ft.	11-7/8"	TJI	blocking
80	shts.	3/4"	birch plywood	subflooring

25	tubes	PL400	Construction adhesive	for subflooring
28	pcs.	JS210	Joist hangers	for 10-1/2" I joist
2100	sq. ft.	4 mil	Black polyethylene	for crawlspace
5000	pcs.	8d	Gun nails	Subfloor nails
1	pcs.	8 x15", 12'	beam	Great Room
1	pcs.	4 1/2x12", 8'	beam	Kitchen opening
WALLS - 1ST LEVEL				
1260	lin. ft.	2 x 6"	SPF	plates
264	pcs.	2 x 6"	SPF, 92-5/8"	studs
120	pcs.	2 x 6"	SPF, 10'	studs
240	lin. ft.	2 x 6"	SPF	blocking, bracing, etc.
780	lin. ft.	2 x 4"	SPF	plates
260	pcs.	2 x 4"	SPF, 104-5/8"	studs
432	lin. ft.	2 x 4"	SPF	blocking, bracing, etc.
156	Linear feet	2 x 6"	SPF	headers
64	lin. ft.	2 x 8"	SPF	headers
98	lin. ft.	2 x 10"	SPF	headers
24	lin. ft.	2 x 12"	SPF	headers
32	lin. ft.	6 x 12"	SPF	headers
124	sheets	7/16"	OSB 4 x 8 panels	sheathing
FLOORS - 2ND LEVEL				
1660	lin. ft.	9-1/2"	TJI	joists
286	lin. ft.	1-1/4" x 9-1/2"	Timberstrand	rims
68	lin. ft.	9-1/2"	TJI	blocking
38	sheets	3/4"	Cedar plywood	subflooring
24	tubes	29	PL 200 adhesive	for subflooring
26	pcs.	JS08	Joist hangers	for 8" I joist

WALLS - 2ND LEVEL				
880	lin. ft.	2 x 6"	SPF	plates
264	pcs.	2 x 6"	SPF, 92-5/8"	studs
168	lin. ft.	2 x 6"	SPF	blocking, bracing, etc.
464	lin. ft.	2 x 4"	SPF	plates
164	pcs.	2 x 4"	SPF, 92-5/8"	studs
260	lin. ft.	2 x 4"	SPF	blocking, bracing, etc.
90	lin. ft.	2 x 6"	SPF	headers
128	lin. ft.	2 x 8"	SPF	headers
88	sheets	7/16"	OSB 4 x 8 plywood	sheathing
FLOORS - 3RD LEVEL / OTHER				
48	pcs.	2 x 10"	DFIR (2), 16'	joists
132	lin. ft.	2 x 10"	DFIR (2)	rims
22	lin. ft.	2 x 10"	DFIR (2)	blocking
1220	lin. ft.	2 x 6"	T&G car decking	subflooring
12	pcs.	LU210	Joist hangers	
WALLS - 3RD LEVEL / OTHER				
466	lin. ft.	2 x 6"	SPF	plates
142	pcs.	2 x 6"	SPF, 92-5/8"	studs
84	lin. ft.	2 x 6"	SPF	blocking/ bracing
80	lin. ft.	2 x 4"	SPF	plates
24	pcs.	2 x 4"	SPF, 92-5/8"	studs
36	lin. ft.	2 x 4"	SPF	blocking/ bracing
48	lin. ft.	2 x 6"	SPF	headers
24	lin. ft.	2 x 8"	SPF	headers
44	sheets	7/16"	OSB 4 x 8 plywood	sheathing
ROOF FRAMING MATERIALS				
1	each		Truss package	
6	pcs.	2 x 6"	DFIR (2), 8'	rafters
8	pcs.	2 x 8"	DFIR (2), 12'	rafters

8	lin. ft.	2 x 8"	DFIR (2)	ridges
12	lin. ft.	2 x 10"	DFIR (2)	ridges
2	pcs.	2 x 6"	DFIR (2), 12'	valley ledgers
2	pcs.	2 x 8"	DFIR (2), 16'	valley ledgers
75	shts.	1/2"	CDX plywood	roof sheathing
33	shts.	7/16"	OSB, LP primed	eave/overhang
4	shts.	7/16"	OSB 4x8	gable sheathing
270	lin. ft.	2 x 6"	cedar	fascia
58	lin. ft.	2 x 8"	cedar	rake
60	lin. ft.	1 x 3"	cedar	rake trim
124	pcs.	H2.5	Hurricane ties	for rafters/ trusses
378	pcs.	H-Clips	Plywood clips	for sheathing
2	pcs.	12 x 18"	Gable vents	
INTERIOR STAIRS				
9	pcs.	2 x 12"	DFIR (2), 18'	stringers
28	pcs.	1-1/8 x 12"	OSB B.N., 3'	treads
10	pcs.	1-1/8 x 12"	OSB B.N., 12'	treads
20	pcs.	1 x 8"	Utility, 10'	risers
12	pcs.	1 x 8"	Utility, 12'	risers
18	pcs.	LU28	Joist hangers	for stringers
WINDOW SCHEDULE				
4	pcs.	5'0 x 5'0	Vinyl, Anderson, white, top awn.	
2	pcs.	4'0 x 5'0	Vinyl, Anderson, white, top awn.	
1	pcs.	3'0 x 5'0	Vinyl, Anderson, white, top awn.	
2	pcs.	3'0 x 2'0	Vinyl, Anderson, white, awning	
2	pcs.	2'0 x 4'0	Vinyl, Anderson, white, fixed	
4	pcs.	5'0 x 5'0	Vinyl, Anderson, white, top awn.	
3	pcs.	3'0 x 5'0	Vinyl, Anderson, white, s.h.	
3	pcs.	3'0 x 5'0	Vinyl, Anderson, white, s.h.	
1	pcs.	3'0 x 2'0	Vinyl, Anderson, white, awning	

DOOR SCHEDULE				
2	pcs.	3'0 x 6'8	Model 8860 dual core door	
1	pcs.	6'0 x 6'8	Entryway	
7	pcs.	2'6 x 6'8	6-panel, hollow core, painted masonite indoor	
2	pcs.	3'0 x 6'8	6-panel, h.c., painted masonite, indoor	
1	pcs.	3'0 x 6'8	6-panel, s.c., painted masonite	
5	pcs.	2'6 x 6'8	6-panel, h.c., painted masonite	
1	pcs.	2'6 x 6'8	6-panel, h.c., painted masonite	
2	pcs.	4'0 x 6'8	6-panel, h.c., painted masonite bi-fold	
2	pcs.	6'0 x 6'8	6-panel, h.c., p.mas. bi-fold	
1	pcs.	16'0 x 7'0	Metal	
SIDING / EXTERIOR TRIM				
7020	lin. ft.	8-1/4"	cedar	siding
1.2	squares		Cedar shingles #275	siding
9000	pcs.	8d	Gun nails	Siding nails
484	lin. ft.	5/4 x 4"	cedar	outside corner boards
92	lin. ft.	2 x 2"	cedar	inside corner boards
326	lin. ft.	1 x 3"	cedar	vertical battens
276	lin. ft.	2 x 6"	cedar	frieze boards
290	lin. ft.	2 x 10"	cedar	belly band
29	pcs.	1-1/2"	Drip edge, 10'	belly flashing
364	lin. ft.	5/4 x 4"	cedar	casing, windows
110	lin. ft.	5/4 x 4"	cedar	apron/casing, windows
110	lin. ft.	1/2 x 2"	cedar	parting bead, windows
115	lin. ft.	2 x 3"	cedar	sill, windows
133	lin. ft.	3/4 x 2-1/4"	cedar	crown mould, windows
60	lin. ft.	5/4 x 4"	cedar	casing, doors

16	lin. ft.	1/2 x 2"	cedar	parting bead, doors
19	lin. ft.	3/4 x 2-1/4"	cedar	crown mould, doors
15	lin. ft.	2 x 10"	cedar	side jamb, gar.door
18	lin. ft.	5/4 x 4"	cedar	head jamb, gar.door
35	lin. ft.	5/4 x 4"	cedar	casing, garage door
15	pcs.	1-1/4"	Shake flashing, 10'	door/window flashing
3000	pcs.	16d	Galv. casing nails	gun nails for ext. trim
DECKS / EXTERIOR STAIRS (ONLY PRESSURE TREATED)				
4	pcs.	2 x 8"	lumber, 12'	ledgers/rims/ blocking
4	pcs.	2 x 8"	lumber, 18'	ledgers/rims/ blocking
12	pcs.	2 x 8"	lumber, 8'	joists
17	pcs.	2 x 8"	lumber, 12'	joists
48	lin. ft.	4 x 4"	post	rail posts
82	lin. ft.	1 x 8"	Cedar	trim boards
784	lin. ft.	5/4 x 4"	cedar	decking
152	pcs.	2 x 2"	Cedar, 3'	pickets
80	lin. ft.	2x6"	Cedar	rail
80	lin. ft.	2x4"	Cedar	rail
160	lin. ft.	1x3"	Cedar	rail
40	pcs.	1-1/2" x 3"	Deck flashing, 10'	deck to wall flashing
42	pcs.	LU28	Joist hanger	for deck joists
300	pcs.	16d	Galvanized nails	gun nails for deck .
2000	pcs.	3-1/2"	Galvanized screws	Deck screws
3	pcs.	2 x 12"	lumber, 12'	stringers

6	pcs.	2 x 6"	lumber, 10'	treads
3	pcs.	LU28	Joist hangers	stringers
18	pcs.	TA10	Staircase angles	treads
140	pcs.	1/4" x 1-1/2"	Galvanized lag screw	angles
6	pcs.	12x12x8"	Conc. w/saddle blocks	Rear deck
INTERIOR TRIM				
880	lin. ft.	1/2 x 2-1/4"	pine	casing, doors
294	lin. ft.	1 x 6"	pine	jamb, windows
364	lin. ft.	1/2 x 2-1/4"	pine	casing, windows
110	lin. ft.	1/2 x 2-1/4"	pine	apron/casing, windows
115	lin. ft.	1 x 6"	pine	sill/stool, windows
68	lin. ft.	3/4 x 5-1/4"	pine	crown moulding
68	lin. ft.	1/2 x 3"	pine	chair rail
961	lin. ft.	1/2 x 2-1/4"	pine	baseboards

Siting the House on the Lot

If you are like me, you thought the long and arduous task of searching for a lot was the time-consuming part. Then, when it is time to site the house, you take even longer to decide! That is because this is one of your first "real" decisions, and because so much depends on the orientation of the house. Will you take advantage of the sun, or the shade? Will the house look awkward? Will you regret it after it is complete?

The site cannot be changed once construction begins, so it is ideal to work with the site itself to create a great looking, functional, cost-effective plan. To do this, several factors must be considered.

The neighborhood

Where are the surrounding houses sited? Setting your home further back, further forward, or at an angle compared to all the others may make it look undesirable. If the homes are fairly close together, be sure that yours blends with the neighborhood.

This applies to functional elements as well: if the neighbors do not have long porches or front-end garages, you might want to think twice about being the sole non-conformer. Your home does not have to be a cookie-cutter copy of the others, but it will look best if it fits in with its surroundings.

The view

For you, as the occupant of the home, the view is the most important deciding factor. A view at the back will almost demand that you turn your plan around, with the "front" designated on your house plans facing the back of the lot. Be sure, if you do this, that the garage is still located closest to the street. The importance of the view will mainly affect the family room and master bedroom, and to a lesser extent, the kitchen and living room. Consider foliage, mountains, trees, rocks, and bodies of water when deciding how to place the home.

If you are not going to turn the plan around to face the view, you will at least want to re-design the windows, so that the biggest windows are on the "view side." This is especially a factor if you are keeping the total square footage of windows to a minimum because of heating or cooling concerns.

The sun and the weather

Depending on the climate of your geographical area, you may want to put the majority of your windows facing south, where the sun can warm your home in the winter. However, if you live in an area that has mild winters and excessively hot summers, you may be more interested in cooling; you will want the glass to face toward the north, so that the sun cannot beat in so relentlessly in the summertime.

Another issue is storms. In hot locations, the winds that come with summer storms can cool the house. Storms in a given locale frequently move in the same direction. That means one side of a house is exposed to more severe weather than the others. The direction varies, but most often the north and west of the house are the bad-weather sides. Hot-weather homeowners will appreciate northern-facing windows. In cooler climes, you can compensate by putting fewer windows on the harsh weather side of the house. You might even locate the garage on the north or west side of the house.

The slope

Few building lots are flat. Most of the time, a lot has at least a gradual slope. This will affect the way that people look at the house, how high or low your first floor will be, and the drainage of the lot. It is important to consider all these issues carefully before siting the home.

On a gradual slope downward from the house to the street, the house may seem bigger than it is. You can also be assured that the water will drain, although maybe not in the right direction. Most homeowners want the water to pour toward the front because storm drains are located in the street. This would carry the water to the public street and off your private property, but if your water flows onto a neighbor, you could end up flooding their property and possibly even being a party to a lawsuit. The way the water moves will become crucial once the house is built, so pay careful attention in the beginning.

On a lot that slants downward from the street to the home, the view may focus the visitor on your roof line, and cause the home to seem smaller than it is. This could affect the appeal of the home if you should ever choose to sell the property. A low-lying home can be a victim of runoff, and may require an expensive drainage system to stay dry. If your house is on a slope, ensure that water moves around both sides of the proposed home site and down the slope. If there is an area that seems blocked, either make plans to excavate, or include drainage plans in your design.

If a site is particularly steep, a contractor might tell you that the easiest way to build will be "cut and fill." The builder cuts into a slope with a bulldozer to create a spot about half the size of the house, then moves the dirt from that area to level out the rest of the house. The advantage to this is a substantial savings on excavation costs; you will only have to cut into half the dirt and you will not have to pay to haul it away.

Cut and fill should be approached with extreme caution because over the long run, the house may not stay stable. Sometimes the filled side will settle, leaving the other side firmly in place. This can crack the house down the center. A better idea is to put the entire home either on the fill or on the cut.

Sometimes when lots are sloped, the ground can be terraced or a retaining wall can be constructed to create a building pad for the home. Retaining walls may be made of steel and concrete, boulders, logs, or railroad ties. They may be placed above or below the building pad. Remember, a retaining wall has to extend at least two feet into the ground for every foot of height above the ground. So a 3-ft-high retaining wall must be 6 ft deep. Otherwise, the land and the house can slip.

Readying the Site for Construction

Before beginning any work, hire a surveyor. Even if you recently purchased the lot, it is important to designate the lot lines by marking out all the borders. Permanent markers installed by the government correspond to the county plat map. Using the site map, the surveyor, often working in tandem with the contractor, marks out the boundaries, setbacks, and underground utilities. Then they mark the corners of the house and the garage, driveways, walkways, and where utilities will run.

If you have hired a surveyor and a contractor to mark the foundation, go back over all the property yourself. Make sure the batter boards or markings have straight lines, square angles, and clear markings. Be sure that the lot corners and setbacks are marked, and measure them yourself so that you know they are correct.

Now is the time to place clear "Danger" signs around the property, and plan where materials and trash will be stored, preferably away from the building site.

Be sure that the surveyor flags the underground utility lines, and make sure they stay marked throughout the building process. Before beginning any work, call and ask the utility company to mark where the utility lines run. If you do not call them, and you end up causing damage to a line during excavation, your fine could be as high as $100,000 plus damages. If any of the utilities are in the way of driveways, landscaping, or other items, they can be relocated.

Site preparation includes roughing in a driveway, cutting down trees, and possibly installing the well and septic system or digging trenches for the utility hookup. Be sure that you get enough trees cut down to make room for the new home, plus the large trucks which will have to maneuver around the foundation once it is in. While this is going on, clear a path for the utility trucks to bring in the electrical service. You can save a little money by making sure you get enough trees removed in the beginning. Otherwise, you will have to pay someone to come back and cut more.

Tree removal

If you have a lot of trees, or if your trees have large diameters, consider trading the wood for the work and getting the lot cleared for free. Some individuals will remove trees to use for firewood. Some logging companies will purchase the timber from you, if it is the right kind and if you have enough of it. However, many of them are scheduled far in advance — I was given a wait time of three years. So investigate this carefully before you make it a firm plan. Also, make sure that anyone who comes on your property to do any type of labor has insurance that covers him or her.

If you plan to burn firewood yourself, talk to the contractor who cuts the wood about having someone chop and stack it on a corner of the lot. If you do not have room to store it all, sell it or give some of it away. The wood is yours to keep if no other arrangements have been made.

Once you the trees are cut down, there will be stumps to remove. These are large, often as tall as four feet above the ground. Do not be tempted to leave them; they will decompose, leaving ugly sinkholes. They must be hauled away and ground up — and sometimes it takes a hefty piece of equipment to remove them. Do not assume that the person who cuts the trees will remove the stumps; you may have to hire another sub to do this.

It is expensive to remove trees, and it is also costly to transplant new trees. Trees take decades to mature, so if you have mature trees already on the property, it makes sense to utilize them, even if it means redesigning the house or locating it differently on the lot than you initially planned. This is taking advantage of natural elements.

To protect trees that you intend to keep on the property, place hay bales around them to be sure that they are not damaged by heavy equipment. Surrounding them with bright orange plastic fence from the hardware store will help drivers avoid hitting them, and protect the trees from flying debris. Alternatively, tie red ribbons around the trees to let workers know not to cut them, and to help them see the trees and avoid hitting them.

Besides tree removal, a bulldozer must come in and clear the shrubs and brush. This is a fairly low-cost activity. After the clearing, it is time for grading. Smaller grading projects can even be done by hand, although tractors and bulldozers more often do this job.

Grading

Grading is the redistribution of dirt from one area to another. Scraping the dirt down creates level ground. Leftover topsoil may be held on the lot until it is needed at the end of the project for landscaping. Finish grading at the end of the project will smooth the yard, as well as ensure that water moves away from the house in a desirable way.

There is not a lot to watch for in grading work, except to be sure that the grading extends three feet or more beyond the edge of the house and garage areas. If the grader discovers lots of boulders, you may have to perform blasting to remove them. Blasting is expensive, and even dangerous, but sometimes it is the only way to remove these rocks.

Occasionally, the rocks are so large that nothing can be done. Several years ago I hired a contractor to bulldoze the back yard. I asked him to dig up or cover over the rocks. There were two of them, each about four to five feet long and three feet wide. They were not a big deal, but they interfered with mowing the yard. He called me outside a few hours later; he had dug alongside one rock and exposed it. Now it was about twenty feet long, and the trench exposed it to a depth of about five feet. The contractor assured me there was a lot more to it. The boulder was much too close to the house for blasting. We covered it back up and did not attempt removal.

How many subs do you need for the site work?

The more contractors you have working on the property, of course, the faster the job will progress. This can save you money. On the other hand if you prefer to have one person do most of the work you can wait for him to finish. Normally, you hire one contractor to remove trees; one to dig the foundation, remove the tree stumps, and rough in the driveway; and possibly a third to install a septic system. The trenches for the water system may be done either by the earthwork contractor or through the plumbing contractor. Schedule carefully for all this; you do not want the septic tank sitting in the way of the grader, or the tree cutters to spill over into the scheduling for the stump removers. This is why some people choose to just use one or two subs for the site work.

Install temporary utilities

Temporary utilities, water and electricity, must be in place before you can do anything else. The subs will need them to do their work. Talk to each utility company well ahead to find out what you need to know.

Public water and wells

If you are connecting to public water, a trench must be dug to the water supply, and the main water pipe connected at the street. A temporary meter must be installed so that the water company can monitor the amount of water used during construction. If the main pipe bringing water in has pressure of greater than 80 psi, it requires a pressure regulator to reduce the water pressure so the pipes are not overly stressed. In the worst case, the pipes may even burst without it.

In a more rural area where public water is not readily available, you will install a well and septic tank. Both of these must be done by licensed professionals. First, the well company will drill down into the water table until they find clean ground water. Once it is located, the workers install an electric pump which brings the water to ground level for use. A water storage tank is installed so that the pump does not have to kick on every time you run a small bit of water inside the house.

Whichever type of water you install, be sure that the pipes do not leak at any location. Make sure all the connections are secured, and be sure that the spigot is flagged with a red ribbon so that no one runs over it.

Sewer and septic systems

Public sewer systems require cast iron or PVC pipes which are laid in trenches before any concrete is poured. The sewer pipes must slope downward toward the sewer system. Be sure that you install an anti-return valve on the pipe, so that if the system overflows (during heavy rainstorms) the water cannot back up into your house. Clearly mark the connection point, and after the pipes are properly buried, have the inspector out to check the system.

If you are installing a septic tank, call a septic company out to choose the best location. The septic system is comprised of a tank and a leach field. The wastewater from the tank drains into the leach field. The more bathrooms and bedrooms you plan to have, the bigger the tank that will be required. Tanks of 1,200 to 1,500 gallons will service a 3-bedroom, 2-bath house. That may sound large, but the tank is buried underground and does not require a lot of room.

The tank does need to be positioned in a convenient spot so that a truck can approach it for cleaning. After the tank is in place, mark it clearly so that it is not covered up during excavation. A septic company must inspect the tank once it is in place.

The problem, if there is one, will be with the leach field. It has to cover enough ground to absorb all the wastewater that the septic system sends into it. This is ultimately dictated by the type of soil. A swampy soil may not be suitable for a leach field; a very wet soil will require a large leach field; and a fertile, friable soil will require a smaller field. You will have to have the soil perc tested to find out what size of field is required.

There will be setbacks from the property line, and these include your septic tank and leach field. Be sure that your lot is big enough to accommodate these and still conform to setback requirements. Hire a waste engineer to plot it out if necessary.

Some localities now require that you have enough room for two leach fields: the primary one that you are planning to use, and a secondary one in case the first one becomes plugged. This is a real possibility, especially if you do not have your tank cleaned every few years. If your locality requires the space for a backup leach field, make sure that your lot is big enough, or you will not be granted the necessary permit.

Electricity

Unless you are going to live "off the grid", you must run electricity into your home. This will either come in underground or from a pole; you can look at the property around you to see which way it is done in your neighborhood. The power company will run the line up to 50 or 60 feet at no charge, but if your site is very far from the pole it may cost as much as $25 per extra foot. You or your subs can do the trenching, if it saves money, and you can have a licensed electrical contractor install a temporary meter, but the utility company will do the connecting. In some areas, the contractor who installs the meter must be on the electric company's approved list. The building inspector will need to inspect your meter before the electric company connects the power.

Public gas and propane

Connecting to a public gas line is simple; the pipes are trenched to the gas line that runs under the street. For propane, you can get the supply company who will be delivering your propane to install the tank. This can wait until later on. It is not necessary to have gas or propane until late in the building process.

Be sure that the gas pipes are properly buried, when the time comes. Clearly mark the connections.

Excavation

Excavation should cover all areas of the foundation, fireplace, porch, and stoop. The crawl space should also be cut and graded in a way that makes sure it stays dry. You absolutely must watch the excavation process. Do not be tempted to let the sub set this up without you. The whole time the work is being done, take depth measurements. If the foundation is too low, the sewer and water lines will have drainage problems; the sewer line could also be lower than its street side connection. Foundations that are lower than the land surrounding them are susceptible to basement water problems. Water lines and footers must be placed well below the frost line, which can be 2 to 5 ft. in the northern parts of the US. If the foundation is too high, you will end up having more front steps. Keep an eye on the work, and if it has to be a little off, let it be a touch higher than planned. You can expect all the excavation work to be within 2 in. of level.

While the excavator is working, have him clear the topsoil from the driveway, and pile all the topsoil where you can get to it later. Be sure that the excavator digs below the frost line, wherever it is. (Confirm where the frost line is for your geographical area.) You can save money by having the water and sewer lines run in the same trench; go ahead and have this trench dug during foundation excavation.

Now install the batter boards, checking the corners for the correct angle as mentioned earlier. Batter boards are boards nailed horizontally to posts set at the corners of an excavation. They indicate the desired level for the foundation. They also serve as a base for fastening tautly stretched strings that mark the perimeter of the building. Here is how to lay them out: Drive three stakes outside each corner post. Use sharp 2X4 stakes; make sure they are long enough to protrude 6 in. or so above the boards. Now, using boards of either 1X4 or 1X6, nail a board horizontally between a pair of stakes, at the top and bottom of the triangle. Repeat the process, placing another board from the bottom stake to the rightward stake. Lay the outside batter boards first, square them, then lay the inside batter boards. These batter boards serve to outline the foundation, leaving room for digging around them.

Drainage

Now it is time to put a run of crusher stone onto the driveway. This will be used as a base for the paved driveway later, and it helps allow suppliers and subs to get in and out now.

The local government will not allow mud from your construction site to run onto the street or other people's property, so build a silt fence to keep the mud in. This is done with 2-ft. stakes and rolls of plastic or mesh. They can also be built with hay bales, which are then re-purposed for landscaping later.

Foundation

The foundation phase is one of the most important parts of building your home. Hire the best contractors you can find. If you can locate a full-service foundation outfit, they will lay out the entire foundation, including the batter boards. They will then dig/pour the footings, pour the basement floor, and schedule all the inspectors. Most of them even do the waterproofing. These companies are a little more expensive than hiring the work separately, but you will save in time and costs, and you might even save some on your construction loan interest. Full-service contractors are worth checking out, especially if your plans call for a full basement.

If you are not going to hire contractors to construct the footers and foundation, take the frost line into account. If you happen to build the footers above the frost line, the weather will cause it to continually expand and contract. This in turn can make the footer move, eventually cracking the footer and/or foundation. The purpose of a footer is to distribute the weight enough so that the house does not settle, and nothing moves.

To pour a concrete foundation wall that will enclose a basement, measure a minimum of 7 ft. from the basement floor. Many contractors go up to 9 ft. When back-filling the foundation, use the soil that was excavated to create the foundation — not the topsoil that was originally piled into a back corner of the lot when you first began construction.

Chapter 10 Checklist

☐	Investigate local suppliers for building materials
☐	Prepare a materials list
☐	Get quotes from various suppliers
☐	Review the siting of the house on the lot considering: • Neighborhood houses • View • Sun and weather • Slope of the lot
☐	Review placement of batter boards for location, squareness and markings.
☐	Place "Danger" signs and tape around the property.
☐	Determine where materials and waste will be stored (preferably a significant distance from home site/work area).
☐	Clearly mark all trees, showing those which stay and those which are to be removed.
☐	Remove trees and brush.
☐	Grade the property.
☐	Install temporary utilities.
☐	Install the septic system.
☐	Excavate as required.
☐	Pour foundation. Build up with block as required by the plan elevation.

CHAPTER 11

First Stages of Construction

Framing

Framing the house is always exciting because it suddenly looks like a great deal of progress has been made. After months of planning and working, it finally looks like a real house. You can see the plans you made on paper taking shape.

The framing of the house is absolutely critical. It is the support structure for the entire project, much like the skeleton of your body. There is no room for error in framing; any deviation over ¼ inch is considered unacceptable.

Framing lumber

The lumber used for framing is kiln-dried; the higher the grade, the better. I have seen people use green, freshly milled lumber for floor framing. As it dried it buckled, creating a wavy floor for the homeowners to walk on — not to mention an unsafe support system for the rest of the house, and a guarantee of shifting and settling later. Even if the lumber is seasoned but lower grade, it sometimes shrinks after it is in place. This will cause "nail popping," which is annoying and requires repairs. The better your lumber, the easier it will be for you or your sub to do a good job. If lumber is going to touch the foundation or concrete, it should be pressure-treated pine, which will protect against moisture and termites.

If you have purchased a kit home, there will be precut materials that are possibly even preassembled. They will have their own set of instructions.

Floors

If your house is built on a concrete or block wall, bolts will be inserted in the concrete as it is poured for attaching a sill plate. The sill plate is a board which is bolted to the foundation and becomes the anchor point for the floor and all exterior walls. Floors are framed to make a base for the floor to rest on. This support has to be strong and stiff, with no give that could result in settling. Joists are made of 2X 8s or 2X10s, and are placed 12, 16 or 24 in. apart.

Floor joists that are excessively long are braced up by supports like steel I-beams. There are also vertical supports made of steel that rest on footings or piers. It is important to consider what the post is resting on; some builders place a wooden block and nail the post to it. Instead, embed steel plates into the concrete footers and bolt the posts on these. This provides strong support and a barrier against termites and other pests. Some contractors suggest using two boards fastened together, creating a double floor joist, around openings or spaces that are created for hot air ducts or bathtubs.

Sometimes instead of traditional joists, the boards are bridged (stabilized by using bracing placed diagonally between the joists). If this is the case, they are called trusses. Trusses are becoming popular, not only because of their strength but because they make it easier to install HVAC and wiring, and because they tend to have better sound-reducing capabilities.

Subfloor

The subfloor is the platform to which you attach the finished flooring and the walls. It is made from plywood or particle board and lies in the opposite direction of the joists. You may choose not to lay subfloors. If you can place the flooring straight onto the slab there will be no need for subflooring. However, the upper floors will still need subflooring, as will a house that is built on a beam foundation.

The subfloor is the main cause of squeaking floors. Two precautions are necessary to prevent squeaks. The first is to glue the plywood down when building the subfloor. The second is to use ring shank nails (the kind with screw threads) to nail down the plywood. These will hold the wood down and will not draw back out and create room for the plywood to flex and squeak. Another alternative is to use joists made of composite instead of wood, so that they are not able to give.

You may have heard of a "floating" subfloor. This is thicker plywood flooring than the traditional ¾ in. It is usually created from 1½-in. plywood squares placed on piers. These are used in areas where there is a lot of expansion and contraction, due to weather, soil, or temperature fluctuations. The floor can move slightly to accommodate the expanding and contracting. The finish flooring rests directly on top of the subfloor. If you have a floating subfloor, you will not notice a difference between it and a more solid floor.

There is one exception to the subfloor style, and that is a subfloor for tile in the bathroom or the kitchen. If you plan to use tile in these spaces, you must use a special compressed wood made especially for tile. It is laid over the plywood, then screwed, glued, and sealed. A cement surface is even better for tile because cement can be leveled and smoothed perfectly. It can be strengthened with wire mesh, then the tile glued directly to the cement.

Walls

Once the subfloor is laid, it is time to build the walls. Wall framing has to be exactly right, so that there are no surprises, no changes, and your house does not look like it was built by the "crooked little man" referred to in the song.

Making changes to the framing can cause many design issues, so consult with the architect one last time before starting the wall framing. There are also some plumbing considerations. The framing should be spaced with room for the plumbing that will be put in. Some shower stalls, especially the one-piece kind, may not fit through a regular-sized doorway once the framing is up. Put the shower stall or Jacuzzi tub in place prior to framing. If you are concerned about theft, secure these in place with heavy chains.

Types of wall framing

Most framing in the United States is platform, or stick, framing, in which all the walls for the first floor are attached to the subfloor. The second floor rests on the ceiling joists for the first floor, and so on.

Another type of framing is balloon framing. In this type of framing, the vertical exterior walls extend from the sill plate on the first floor to the top plate of the top floor. These studs are more costly than the studs used for stick framing. The labor is also slightly more costly for the construction. Balloon framing requires that the entire outer perimeter of the structure be uniform, so it is not always the best kind of framing for a given set of plans.

Pole framing can be used when strong winds are expected. Although it is not a common kind of framing, the poles are sunk deep into the ground and are exceptionally secure.

Post and beam framing uses rough-hewn wood to span long, open spaces. It does not use joists for the floors or the ceiling. This kind of framing is used for exposed beams and open ceilings.

Steel post construction is quickly becoming a standard in home construction; steel posts are used at the corners of the structure, embedded in concrete. This is used in conjunction with walls that are filled in and do not support the ceiling above. For houses that are exposed to strong winds, metal hooks can be placed in the foundation with bolts that fasten to the steel rods; if the house moves up and down, the wall will not separate from the foundation. Some builders run a metal brace on the top plate around the entire perimeter of the house as well, attaching it to the corner supports.

Money-saving Tip: Bracing the corners of the outer walls

The corners of the outer walls are required (by the local building code) to be braced. Using a diagonal wood brace cut into the studs, or a metal strap, which will save money and will allow for the insulating that is created by the sheathing.

Homes in the US are built with 2X4 framing placed 16 in. apart. Stud grade lumber will cost more than standard framing lumber, but the studs will be uniform and straight, and will not have as many cracks. Whichever one you choose, do not be tempted to frame with "utility" 2X4s. These are uneven, making it nearly impossible to finish the walls properly, and they may not even support the load required.

Be sure that you have made plans for fire blocks to fit between the studs, midway between the floor and the ceiling. The function of a fire block is to keep fire from moving up through the walls to the next floor.

Before creating the framing, make sure that the placement of any openings will allow for good airflow. Better airflow takes place when air comes in and out through openings (windows) of about the same size. Make sure the windows are located in a way that allows for cross ventilation. Even though many homeowners today never open the windows, there will be times when you will want to air out the house, dry freshly cleaned carpet, and so on. You will be glad to have good ventilation at those times.

Interior walls

The exterior walls support the weight of the roof. The interior walls, on the other hand, are created for partitioning the rooms and supporting the floor above. These walls are built almost the same way as the outer walls; there are bottom plates, top plates, and studs spaced 16 in. apart. The biggest challenge in interior wall construction is soundproofing the walls. Normal construction (2X4s with wallboard nailed over them) creates a hollow cavity where sound echoes and transmits easily to the next room. Since you are building or remodeling your own space, it is worthwhile to insulate the walls to cut down on the sound transmission. The cost for insulating the entire home is only a few hundred dollars.

To insulate the walls, one partial solution is to stagger the 2X4 studs so that each side of the wall has studs that are for that side only; in other words, if we are working on an interior wall that faces the living room on one side and a bedroom on the other, every other stud will be positioned against the living room wall, with the others against the bedroom. After this is done, there is still a need for an additional sound trap. This can be accomplished by filling the wall with insulation. Cut the insulation to fit exactly; do not fold it over inside the wall.

Windows and doors

It is important that windows and doors be framed properly, with double studs and headers that run across the top. The header's function is to support the weight above so that there is no sagging from overhead. Header sizes must be calculated carefully; it often takes strong beams to support all the weight.

Sheathing

Exterior sheathing is made of plywood or particleboard. It creates a weatherproof base on which you will place the outer siding, brick, shingles, or other type of wall materials. If you are going to use plywood, be sure it is exterior grade, which is durable and will not weaken because of exposure to the elements.

Money-saving Tip: Use particleboard for exterior sheathing.

Plywood has become more expensive than ever, so consider using particleboard for houses that will have another exterior covering over the sheathing. If you are going to use particleboard, nail it twice as close as you would plywood.

Roof Framing

The roof protects you and your belongings from the elements. If it is constructed properly, it will keep cool air and warm air where they belong, depending on the season. The roof needs to be sturdy; it has to support tile or slate, solar panels, skylights, and contractors walking on it, and snow loads if you are in an area which gets snow. It also has to allow for items that can cause moisture to enter the home, like vents, skylights, valleys or dormers, and chimneys.

Most roofs are created in one of a few styles: gabled, gambrel, flat, hip, mansard, and shed. The steepness of the roof is called the "pitch." Pitch is stated as the amount of rise (vertical distance) in a given amount of run (horizontal distance). Pitch is described with the rise first: a 2-12 pitch means two units of rise for every 12 units of run. Normally the unit of measure is inches.

PITCH	MEASUREMENTS
Low	1-12 or 2-12
Medium	3-12 to 6-12
Steep	7-12 to 12-12

The pitch of your roof will depend on the amount of snow in your area, what kind of materials you are using for the construction, and how much space is beneath the roof. Climate and architectural appeal are also part of the decision process. A low pitch offers a roof that will shed water, but can hold snow, which adds extra insulation, in cold climates. In the space directly beneath the roof, there will be an area that feels open and has plenty of room for storage, with a sloped ceiling above. Interior maintenance may be easier with a low-pitched roof. This type of roof is less costly than those with steeper pitches, because it requires fewer materials. Low-pitch roofs can be constructed so that you can hardly see them, which might eliminate the need for costly colored roofing tiles. Lower pitches also create a smaller overall cubic footage (length X width X height), making that the cost of heating and cooling lower.

Roof framing is either stick-built or pre-manufactured roof trusses. The trusses are extremely sturdy. They can hold up a large load and they do not require that there be load-bearing walls placed above the exterior walls, so there is more design flexibility. They can be installed in a single day. Roof trusses are also less expensive than stick-built roofing, because there is a savings in labor costs. The negative aspect of roof trusses is that you will not have as much attic space to use for storage.

If you are creating a cathedral ceiling, you will probably use rafters instead of trusses. These are boards that cover the entire distance from the exterior walls to the peak of the roof, forming a "skeleton" framework on which the roof decking is attached. Rafter sizes are computed based on load-bearing qualities of the wood and boards used, the distance that must be spanned, and the pitch of the roof. In addition the spacing, lumber grade, local wind forces, and amount of yearly snowfall is a part of the computation. One of the best choices for rafter joists are engineered wood I-beam roof rafters. These are attached with special blocks. They are unusually strong and can span greater distances than most wood. The dimensions are precise, which means no time is wasted adjusting for size and there is no wasted material. They do not shrink or crack, and are lighter than most lumber. Like all rafter roof framing, the load-bearing interior walls are used for support.

Roof Sheathing

The roof sheathing is made of pressed chipboard or ½-in. CDX plywood. Be sure to place plywood clips between each sheet to keep the roof's surface smooth.

Roofing Materials

The choice of roofing material affects the style and the look of your home. Let us look at some of the options.

Wooden shingles (shakes)

If you plan to use wooden shingles instead of typical asphalt or composite shingles, place furring strips on the roof spaced with 3 in. in between. Cover the boards with roofing paper, then lay the shingles on top. The shingles will be able to breathe, so they are less apt to become covered in mildew. Wood shingles can be a fire hazard, so they are used less than many other types of shingles. They are also more expensive to purchase and install.

Ceramic or clay tiles

These are an especially attractive option for a Spanish or stucco house. Unfortunately, they are costly. They also may be extremely heavy, up to 75 percent heavier than metal roofing. Consider the pitch of the roof and the load-bearing capabilities, especially if you are re-roofing an older home.

Asphalt Shingles

These are the traditional shingles used on homes. They are somewhat heavy, but they cost less than all other types of shingles. Most asphalt shingles carry a 15- to 25-year warranty.

Fiberglass and Composite Shingles

These look similar to asphalt. They are flame-retardant, and have a long life. They come in a self-sealing version, which allows the heat to "glue" each shingle to the one below it over time.

Metal roofing

Metal roofing is available in shingles and in sheets. It comes in many colors, is anodized, and it lasts longer than any other shingle. Most metal roofs are guaranteed for 30 to 50 years, compared with 15 to 20 for a traditional asphalt roof. However, a metal roof can dent if someone walks on it, or if a tree branch hits it. There are several types of metal to choose from; aluminum is very lightweight, but must be coated to give it a proper finish. Steel roofs are normally coated in zinc or a combination of zinc and aluminum to prevent rust, then covered in an acrylic topcoat, which adds color and protection from the elements. They are the sturdiest option.

Copper roofing is expensive, but it weathers beautifully. Copper is soft and easy to work with — but dents easily.

Saving Money on Roofing

To save the most on roofing, be sure that the system you choose works well with the elements in your area. If your state tends to have lots of hail storms, for example, choose a roof that has passed impact resistance tests. Check with your insurance company. Many insurance companies offer a premium discount if you select certain types of roofing materials.

Look at the heat resistance and cooling capacity of your roofing materials. A light-colored metal reflects heat, which lowers your utility bills.

Choose a roof system that uses concealed fasteners made from a metal that is compatible with that of the roof. Some roofing systems use clips to fasten the panels together, rather than methods that put holes through the panels. Anywhere there is a hole, there is potential for water to enter the roof and ruin your home.

If you are remodeling and installing a new roof, consider using metal. It can usually be laid directly over the old roof system, eliminating a lot of time and waste disposal.

Roof Overhang

Although it may be tempting not to create much overhang in order to save on materials, there are several good reasons to create an overhang. It blocks the sun when it is high overhead, keeping the house cooler. In winter, an overhang allows the sun's rays in (because the sun is lower in winter) to help heat the house.

The roof overhang also protects the sheathing, siding, doors, and windows from the weather. An overhang at an entry protects you from precipitation as you go in and out. An overhang above the window will allow you to enjoy the sound of rain falling, even with the windows open, without the concern that the rain may come inside.

Large overhangs are associated with fewer moisture problems, both on the exterior and on the walls of the foundation. "Wet walls" can cause mold and even lead to the rotting of wall studs. If your area has significant rainfall, use overhangs liberally. The optimum length of the overhang from the wall has to be calculated based on the angle of the sun in summer and winter—note that it is different at different times of the year — as well as the rainfall. Rain that is falling in winds of 30 mph may fall at a 53 degree angle; if the wind is only 10 mph, it falls at 22 degrees. The typical rains and direction of the prevailing winds in your locale will have much to do with the way you create overhangs.

Flashing

Flashing is the material used on roofs around the chimney, vents, roof valleys, and any other joints or areas where different materials come together. Flashing protects the framing and insulation from the effects of weather. It should be applied in a way that causes the water to flow off the roof, and not pool or pass through to the underlayment. If your flashing is inadequate, or if it is not applied properly, it can reduce the energy efficiency of your home and allow damage to the ceilings and walls.

When most people think of flashing, they are referring to aluminum. It is the most common material used, mainly because it is inexpensive and lightweight. Aluminum is also fairly resistant to corrosion. It is shiny, though, and it is hard to paint, so aluminum is not the most attractive protective coating to use if much of the roof is visible. Copper is another option; it weathers to a lovely green patina, but it is more expensive.

If aluminum flashing is unattractive, vinyl is a good second choice. Vinyl materials come in many colors and are not expensive. Since many soffits are also constructed from vinyl-coated sheets, it is possible to have a smooth look by incorporating vinyl — and you will have almost no maintenance.

Chimneys

Flashing around the chimney is especially important, because the chimney is frequently a source of water damage. Water seepage can damage the roof sheathing and framing. This often occurs before you realize there is a problem with the chimney flashing. The flashing that is placed between the roof shingles and the chimney sides is meant to keep the area watertight.

To ensure the effectiveness of the chimney seal, use two layers of flashing. The first layer starts at the lower side of the chimney, working from the roof line upward. Some contractors take an extra step, wrapping these pieces into the shingles to create a woven solid layer. Over this layer, add a second layer from top to bottom, arranging it so that it covers the top of the snug layer underneath. Seal the entire installation with high-temperature silicone caulk, especially the corners. Do not use spray foam-type caulk against the chimney. When the weather is bad or the roof shrinks, the double layer will keep the area watertight.

Occasionally a chimney is located in the line of fire — right at the point where water is diverted from the roof. If your design involves this type of chimney design, consider creating a saddle from sheet metal that is shaped like an inverted "v." Install it against the chimney and seal it carefully, flashing at all joints. In addition to diverting water, it will keep debris and snow from piling up behind the chimney. This is especially helpful if you cannot see the chimney from the ground.

Gutters and Drainage

If water pools on the roof, it will work its way under the shingles. Water needs to be diverted quickly, and in the most efficient way. To accomplish this, flashing will be added to channel the water down toward the edges. Gutters will then carry the water away from the house. If you allow the water to run directly from the roof to the ground, it will probably seep into the foundation of the house — giving you either a foundation problem or a basement flood.

Guttering is made from aluminum, fiberglass, copper, or plastic. It channels water to drainpipes that are placed at each corner of the house. The guttering must slope down toward the downspout at a pitch of 1 in. per ft., as well as slant slightly from the house so that, even if the gutters overflow, the water will spill outward and not onto the walls. There should be enough downspouts to effectively drain the gutter in a hard storm. The drainpipes carry the water either to the storm drain, or just away from the house.

When choosing gutter, take into account the amount of roof that drains toward each one. "Normal" gutters are 5-in., but it is often necessary to increase a gutter to 6-in. to fully collect the volume of rain. This is especially true for broad, steep roofs.

If you have a tall design that requires getting up on ladders to clean the gutters, or if you have many feet of gutter — or if you simply do not want to climb up there to clean them, consider installing gutters with a built-in gutter guard. The guard keeps leaves and debris out, while allowing water to run in. These gutters cost more, but might save in maintenance costs.

If gutter guards are not an option, consider installing gutters that have removable screens across the top, or removable caps on the ends. These are easy to take off if you need to hose out the gutter with a garden hose.

Downspouts come in several shapes: round, rectangular, and various corrugated styles. The shape does not matter, but the corrugated kind handles freezing water more easily.

Ventilating the Roof

Ventilation is one of the most important aspects of building a roof. Effective ventilation is necessary for many reasons: to remove odors and moisture, to remove hot air from the attic space, to cut cooling costs, and even to extend the life of the shingles. Without proper ventilation, the heat in the attic can literally cook the shingles. Good ventilation will also help prevent the formation of ice "dams" on the guttering and the roof channels.

If your plans call for an attic, you can simply install a vent at each end of the house to allow moisture that accumulates in the roof to dry out. If you do not ventilate the attic, you will find that the temperature can become unbearably hot. Consider installing fans with automatic thermostats in the vents. These are inexpensive, and they will prevent the house from overheating in the summer and extend the life of the roof. You should also have vents in the soffits to provide air flow from the overhang, up the roof, to the end vents.

If there is no attic, use ridge vents to ventilate the roof. Normally a 1-in. gap is left in the plywood right at the ridge of the roof; the vent is installed here.

Exterior Walls

The part of your home design that is noticed the most is the exterior wall materials and finish. Many people describe their house by its exterior: "the painted beige house," or "the brick two-story." Thus, it is important to use a good scale, as well as pleasing textures and colors, on the outside of your home.

In addition to its cosmetic appearance, the exterior of the home offers one more protective layer from the elements, noises, and sudden temperature changes. Because of the expanse of the exterior, you have an opportunity to save thousands on your home's siding. Take into account the cost of the materials, installation, and labor. Sidings are available in numerous thicknesses, finishes, and colors. Some are installed horizontally and some install vertically. Many types of sidings, if combined together correctly, not only look great but also eliminate much of your outside maintenance.

Besides the look and durability, there are other points to consider when selecting siding:

- Weather resistance — whether it will fade or spot over time, and its ability to bend during pressure from wind, hail, and snow, and moisture resistance
- Durability — resistance to scratching, chemicals, and to stray balls, tree limbs, and other objects that may strike it. To what extent do these threats exist around your home?
- Fire resistance
- Sound absorption
- Insect resistance

Siding is a very personal choice, depending on the look you have in mind, and there are so many to choose from.

Aluminum Siding

The Pros:

- Aluminum is a low-maintenance wall covering. It has a baked-on finish and does not warp or crack. It is manufactured in many finishes or textures. Some aluminum siding is designed to look like wood —

painted wood, shakes, weathered barn wood, and so on. A thicker gage of aluminum gives you a stiffer siding that will sustain more wind and hard strikes from flying objects.

- Aluminum panels are offered in both horizontal applications, which might have a width of 4 to 8 in. like clapboard, or there vertical sheets that come in 10 in.- to 16-in.-wide strips.

The Cons:

- Aluminum siding can dent if it is struck by an object. If you have kids who play ball, or neighbors who do, back the aluminum with Styrofoam backing. Also, striking the surface can scratch off the finish so that the shiny aluminum beneath is exposed — hardly an attractive look. The price of aluminum siding is based directly on the current cost of aluminum, which can vary wildly.

Vinyl Siding

The Pros:

- Vinyl is popular because it is low-maintenance, strong and durable. It is also molded with color throughout, so scratching it does not cause a color change. To make it stronger and increase the insulating properties, it can also be backed with the polystyrene board, like the aluminum siding. Vinyl siding does not dent and can withstand powerful winds.

The Cons:

- Vinyl siding can easily buckle if the installation is done incorrectly. I saw a house with brand new siding — and ripples along every panel. The young, inexperienced crew the homeowner hired was not familiar with its use. Vinyl may also be difficult to install around unusual wood trim. It requires careful caulking at every seam, especially where it joins with other materials. Also, if aluminum (non-rusting) nails are not used, there will be rust smudges on the siding in a few years.

Wood Siding

If you want a rustic appearance, consider using plank siding, like redwood, fir, pine, hemlock, cedar, spruce, or cypress. Solid wood siding is offered in many popular styles: beveled and beaded, v-groove, tongue-and-groove, and many other variations. Even the old-style clapboard siding is created from wood.

It is also possible to use plywood panels for siding. These are offered in 8 to 12-ft. lengths and are very easy to install, which can be quite a savings. Plywood can also be applied over the wall studs without underlayment — a further cost reduction.

"Hardboard" sidings are panels manufactured from leftover wood. They are stiff and look like authentic wood. These sidings have proven to be more dense than wood, and they do not split off into layers or crack. They can also be ordered already primed for the paint finish.

The Pros:

- Wood siding is inexpensive and gives character to a plain design.

The Cons:

- Every wood siding must be treated with weatherproofing. The coating will have to be repeated occasionally to prevent water damage, insect infestation, and erosion. This is true even of the pre-finished hardboard sidings. Wood also requires that a vapor barrier be installed underneath, to prevent condensation.

Brick and Stone

Brick or stone exteriors, or a combination of both, are beautiful finishes. They are durable and offer a positive selling point, as well as a good return on investment when you decide to sell the home.

The Pros:

- Brick and stone homes are sturdier when strong storms come; they do not sway in strong winds and will not crack. They are fire resistant and naturally repel a great deal of noise. They require the least maintenance of all wall coverings.

The Cons:

- Changing the exterior can prove to be a huge expense. It will be difficult to match the brick far down the road, and if you decide to remove a wall it can be quite costly.

- Brick and stone are not strong insulators. Frequently, the homes covered in these materials are the hardest to heat. To offset this, you will have to use lots of insulation.

- Because of the thicknesses of these coverings, the exterior walls should probably be moved inward 5 in. to accommodate their thickness. Also the doors and window openings will have to be reinforced with stone or metal lintels so that they can hold up the brick above.

Stucco

Stucco is a type of Portland cement that is applied with a trowel directly over the masonry or on a vapor barrier called stucco wrap. Stucco is applied in at least 3 coats totaling less than 1 in. in thickness. Although it has gotten a bad name in some areas of the US because of poor application, if done correctly, stucco is a sturdy and useful wall covering. Many home buyers feel that the "look" of stucco gives more value to the house.

The Pros:

- Stucco is easy to apply and lasts practically forever. It does not require maintenance unless mildew attacks; this can easily be removed with a sprayer filled with a combination of household cleaner and bleach. Stucco can be tinted with color so that it does not have to be painted, and the final coat can be done in one of several textured finishes, depending on whether you are creating a traditional or a more modern architectural style. If the stucco does crack, it is easy for a do-it-yourselfer to repair.

The Cons:

- On a wood-framed structure, it is imperative that a layer of stucco wrap (a type of construction fabric) is used. The wrap prevents moisture damage in the stucco by routing the water away from windows, doors, and joining areas. If this is not applied correctly, the stucco can crack. It will also crack if there are large walls without control joints placed at a distance of 3 ft.

Wood shingles or shakes

Wood shingles or shakes can be made from redwood, cedar, or cypress. They give a warm, rustic appeal to a home. Cedar darkens and weathers to a silvery color that many people find particularly attractive.

When buying shingles or shakes, you will see that the shingles have two smooth sides, while the shakes have at least one rough side. This is the difference in the two. Higher grade shingles and shakes will not have knots in the wood or pockets of pitch, which can interfere with application.

The Pros:

- If you are trying to create a rustic or woodsy look, you cannot go wrong with wood shingles. Wood is a good insulator, and in some climates can be installed with no preservatives.

The Cons:

- Most wood shingles and shakes must be weatherproofed every five years. This can become a considerable expense over time. Wood shingles and shakes can be expensive, and often they are formed separately so that you have to nail them on one at a time. To prevent ruining their look, use strong, non-rusting nails.

- To avoid some of the expense and time involved in covering your home with wood shingles, look for prefabricated 8-ft. panels, or consider the vinyl or polypropylene versions of "wood" shingles. These panels stand up to the elements and are resistant to warping and twisting.

Exterior Finishing Checklist

☐	Select materials for the house exterior considering: • Style of house • Neighboring homes • Material durability, weather resistance, insect resistance • Cost • Aesthetic qualities
☐	Install exterior materials
☐	Caulk around all windows, doors, other openings
☐	Paint or stain as required

Plumbing

Plumbing System Design

For new construction, planning for the plumbing in your house is a critical task. Plumbing could be the most difficult single item to change once the house is complete. (Some builders argue that the HVAC distribution system is the most difficult to change.) Therefore, plumbing layout needs to be carefully thought out, drawn out, and then thought out again. If you are doing a remodeling project in a bath or kitchen, the plumbing modification can be relatively simple or very difficult, depending on the extent of the project. The addition of a master bedroom suite with a bath will require tying in to the existing plumbing and can present some difficulties. Remodeling is discussed further in the section on inside piping.

When laying out the house floor plan, bathroom placement deserves some special consideration. Placing two bathrooms back to back so that they can share supply and sewer piping can save significant cost. The same thing is true of putting a second floor bath over a lower floor bath or kitchen. Also, one of the most commonly overlooked needs in plumbing design is outside faucets. Think about your needs for water outside, and plan enough convenient faucets to serve all them. If you are going to have an unfinished basement, plan ahead for adding a full or half bath in the future and rough it in. This will cost very little now and will avoid expense later if you decide you need one. It can even increase your resale value by giving future owners an option. As you think of the overall plan for making water available in your house at all the desired locations, there are three main components you need to consider:

- Water supply to the house
- Interior piping to the desired locations
- The waste system and connection to a sewer line or septic system.

We will look at these three subsystems separately.

Water supply

This has already been discussed in Chapter 10, but it is a major consideration which warrants another mention. The water supply to the house may come from a public utility or a private well. Most builders choose to use the public water supply if one is readily available, simply because it is the most convenient option and the one which leaves the responsibility of maintaining the operation and ensuring the purity of the water to others. However, in most areas, drilling a private might be an allowable option and, if you plan to live in the house 10 or more years, could save money over time. The decision is one of personal preference and economics.

To make the financial determination you will need information on the average depth of the water table in your area, drilling costs, water quality, and the required treatment. Between the Health Department and the drilling companies, this information should be readily available. You also need an estimate of your water usage to determine what your monthly public utility bill would be. Do not forget to include usage for lawn and garden watering. In addition, with a well, you will typically need to build an enclosure to protect the pump and piping from freezing, and for storing treatment and testing materials. Well systems will have to be inspected and approved periodically by the local Health Department.

Interior piping

The interior piping for new construction and additions is done in two or three stages:

1. Rough-in
2. Tub set and water pipes, often done at the same time as rough-in
3. Trim out

Rough-in includes locating and installing the flanges and sewer lines for all toilet fixtures, as well as for the main sewer lines and branch lines within the house. If concrete is going to be poured for the first level or basement and sewer lines are to be encased in the concrete, the pipes must be put in place before pouring the concrete. All sewer and drain lines must be vented properly to allow smooth water flow. If this is not done, you will get the same effect as turning a gallon jug of water upside down — a lot of gurgling and chugging, but not much water flow. The venting is normally done through a pipe tree (or standpipe) which accepts drain lines from several sources, connects to the sewer line below, and vents up through the roof. These are typically constructed by the plumber depending on the configuration of the sources.

There are some hard and fast rules. Every drain line or sewer connection must be vented to the outside through a standpipe with no obstructions. Each line connection to the standpipe must have a water seal between the vent pipe and the source. (This is the 'S' shaped pipe you see under your kitchen and bathroom sinks). Do not allow a sewer standpipe to vent into the attic because you will not like the way vent pipes break the roof line of your house. You will get, at best, excessive humidity with accompanying mold and mildew in the attic, and probably a house that has a distinctive odor about it. Sewer and septic systems generate noxious and flammable gases during the decomposition process. These gases need to escape and the vent lines are that escape route. If you fail to install a water seal (trap), or if you terminate the vent in the attic, your house will become the recipient of these unsafe and unpleasant gases.

If there is a second story with a bath, the sewer lines will have to be routed down through a wall. The vent lines throughout the house must go up through the roof through a wall. Therefore, these lines cannot be installed until the walls are framed in. The water lines also have to go through these same walls. For this reason, it is not unusual for the rough-in and water lines to be done at the same time, immediately after the framing is done. The tub unit, especially a one piece shower, will often be set into place even before the interior framing is complete, just for ease of movement through the house. Since interior sewer lines from upstairs sources have to come down through the walls, some extra consideration must be given to noise. Who has not heard that middle-of-the-night gurgle from someone upstairs flushing a toilet as the water runs down through the pipe? Fifty years ago all the interior sewer lines were made of cast iron. Since then, some much more economical and easier-to-install materials are being used (more on materials later). Cast iron is very heavy and does not transmit the sound nearly as much as the modern materials. To reduce sound, avoid dropping the sewer through a bedroom wall where possible. If this is not possible, some insulation packed around the sewer line inside the wall will offer some sound reduction. All water lines must be run and stubbed out before sheetrock or paneling is done.

Since all the sewer or waste water flows by gravity, the interior sewer and drain lines must be sloped toward the point where they leave the house. The rule of thumb is ¼ in. per ft. If this rule is not followed, there will be trouble. No cheating is allowed here. Sewer pipe hung between the floor joists must be properly supported and sloped. Be sure to check the piping manufacturers' recommended minimum distance between supports for the specific diameter and schedule pipe you are using. In addition, while water lines can be put through holes drilled in the floor joists, sewer lines cannot pass through a floor joist. The maximum size hole drilled in a joist is 1/3 the size of the joist and all holes must be a minimum of 2 in. from the top and bottom edge of the joist.

Trim out is the final stage. It includes setting of toilets and installation of faucets after sinks, lavatories and countertops are in place. Tiling work has to be complete in showers and baths. The supply water and sewer connections should be complete before the trim out is done so that the system can be tested as areas are completed.

Pressure reduction

This is a good time to talk a bit more about the pressure reduction valve (PRV). It is a very good idea to locate it an easily accessible location (part of the plumbing design). PRVs need readjustment occasionally and will sometimes get clogged with small particulate matter from the water lines. Many people like to locate the takeoff line for the outdoor faucets ahead of the pressure reducing valve to take advantage of the higher pressure for outdoor use. This is fine so long as the pressure does not exceed about 110 psi. Much more than this, and you will find your garden hose life significantly reduced. Water system pressure exceeding 120 psi could cause another problem. The water piping most often used to connect your house to the water system is available with ratings of 100 psi and 125 psi. Depending on which you select, you may have no choice but to put a PRV immediately after the water meter. You would still have the option of increasing pressure for outside faucets by adding a second PRV inside the house, after the take off for the outside faucets. You could then set the PRV at the meter at 100-110 psi and the inside PRV at 80 psi. PRVs are relatively inexpensive and the increased pressure outside is well worth the cost. Most hardware stores, particularly the "superstores," sell a relatively inexpensive water pressure gage which screws onto an outside faucet or clothes washer faucet to allow you to set the PRV correctly.

Sewer system

The third component is the sewer system. Chapter 10 has a fairly complete discussion of the requirements of a sewer system. If you are not able to connect to a public sewer system and must install a septic system, obtain an approval on the size and location requirements early in the process, before you finalize the layout and grading plans for the house. Sometimes the percolation tests turn up unexpected rock and force the placement to a less than optimum location. It is best to discover this during pre-construction, not after the foundation is poured.

All sewer systems depend on gravity. Waste water must flow from the drains and toilets in the house to the sewer line or to the septic tank, and then to the leach field. There are situations where gravity needs a little help. Some houses are located at an elevation below the sewer line or below the required location of the septic system. In these cases, you will have to install a sump below the house elevation with a pumping system to pump the waste up to the sewer line or septic system. These pumping systems are readily available at builders' supply stores, and are reasonably priced.

As you can see, the water supply and sewage systems for you house can have a major impact on house placement on the lot, landscaping, and cost. Proper planning and timely approvals avoid costly mistakes.

Natural gas and propane piping

Fuel piping deserves special attention. Piping for natural gas and propane, as well as its installation, is very strictly governed by building codes. The grade and thickness of materials used (normally black iron pipe and copper tubing) is specified, as well as precisely where and how each is used. This work must be done by the gas company and/or a licensed installer, usually a combination of the two. It will be done under the close scrutiny of an inspector, who will witness a leak test on the system and many other details. A mistake in these systems can cause the total loss of your house and can kill. Leave it to the experts.

Saving money on plumbing

There are a number of opportunities to save money on your plumbing. Some involve material selection, which can save on building costs in terms of material and labor. Some decisions can save you money over time. The most expensive monthly cost related to a plumbing system is hot water (this is an energy cost as well). The first place to look for savings is the hot water heater. Whether you have a gas or electric heater, its sole purpose is to heat water to a predetermined temperature, hold it there until needed, then repeat the process. Water heater efficiency has been improved substantially over the last five years. All water heaters have energy guide labels which will help you pick the most efficient. The labels also contain information on the family size rating for the different heater sizes. Do not buy a larger heater than you need. Look also for heaters that allow you to program a heating cycle, allowing a lower temperature during the day but having the water hot for you when you get home.

Consider water heater placement in your design. Try to locate the heater as close as possible to the major usage areas such as the laundry, kitchen and bathrooms, perhaps at a central location rather than a garage on the opposite end of the house. Be sure your plumbing plan includes a floor drain near the heater, or if it is in an interior location, put it in a pan (available from the same sheet metal shop that does your ductwork) with a drain in it. This is an inexpensive insurance policy should the pressure/temperature relief valve or any of the connections leak unnoticed. One of the biggest uses of hot water is going into the kitchen and turning it on and letting it run until it gets hot when only a small amount is needed. An instantaneous hot water heater which mounts under the kitchen sink gives you up to 60 cups per hour of hot water.

When selecting your faucets and shower heads, new low-flow designs reduce the amount of water used by these devices, while still allowing them to perform their desired functions adequately. Low-flow toilets are the only option available now, so selection of these can be based on more aesthetic qualities.

Materials

The first thing most people think of when they hear the word "plumbing" is pipe or tubing. When determining what materials you want to use for piping, the determining factors are material cost, installation cost, health considerations (approved for drinking water), life expectancy, and pressure rating.

As a do-it-yourselfer, you may be confused by the varieties of pipe available on the market today. Personal preference is always a factor, but many new, improved materials are available. Each type of pipe has a particular use, which often is dictated by plumbing codes in your locale. Climate is also a factor; the pipes used in a temperate zone are different from those needed in a zone that has days of below zero weather. Health hazards in the home are also a concern. This section explains the types of pipe and tubing materials available, and their applications. There is no universally correct answer; different applications call for different types of pipe.

Materials Available

These are the plumbing pipe and tubing materials most commonly used today:

- Rigid copper
- Flexible copper
- Rigid plastic PVC (polyvinyl chloride)

- CPVC (chlorinated polyvinyl chloride
- ABS (acrylonitrile butadiene styrene)
- PEX (crosslinked polyethylene)
- Polyethylene
- Galvanized steel
- Braided connectors

Each type has its own special attributes and usage.

Copper has been the number one choice of plumbers for decades and is probably still their choice for main water lines to the house, hot and cold water supply lines in the house, and running water supply lines and connections under sinks in the bathrooms and kitchen. Coming in two types, rigid and flexible (rolls), it is a versatile material. The connection and joining method depends on the type of copper. There are two basic designations — Rigid and Soft. The following considerations should be a part of the decision making process before deciding on copper.

Before choosing copper, remember that prices for copper have been rapidly rising. Prices have risen so high that people have recently been arrested for stealing high voltage wiring (in service) to sell for scrap. The high cost of the raw material has quickly been reflected in the price of copper pipe and tubing.

Freezing of the water pipes is a significant concern in house design if you live in a cold climate. While the choice of material will not determine whether the freezing occurs, different materials react differently to the freezing process. When water freezes, it expands. Even rigid copper is somewhat malleable (which means that it can be deformed to some extent). Unless the deformity is severe, it can usually expand and survive the first freezing without bursting. However, copper work hardens, as do most other metals. To demonstrate this, take a paper clip and bend it back and forth. The first several times it bends easily, but with repetition it gets more difficult until the metal hardens, becomes brittle, and then breaks. Some of the other types of materials will withstand repeated freezing better than copper.

When putting in the main supply line to the house, the pipe is necessarily buried in the ground below the frost line. If the soil is rocky or has a quantity of small rocks or pebbles, these will rub against the surface of the copper pipe as the pipe expands and contracts, eventually wearing holes and leading to leaks. If your soil is likely to have this issue, seriously consider an alternative to copper.

Rigid copper has been widely used since the 1960s in home construction, and for many years before that in industrial applications. It comes in straight lengths (joints) from 5 ft. to 12 ft. and in various diameters and wall thickness, and is suitable for pressure. Rigid copper pipe cannot be bent to go around corners so all fittings (ells, tees, caps and threaded connectors) are normally soldered to pre-cut lengths of pipe. Soldering, or sweating, is a technique of joining two pieces of metal using a low melting point (180-190 degrees F) fusible metal alloy. For years the most popular alloy for copper soldering was a tin-lead mixture. Since 1988, lead has been banned in drinking water applications and normally a tin-antimony or silver alloy is used.

Soldering technique can be quickly mastered by the do-it-yourselfer. The secret is to keep the two pieces being soldered be clean and dry, and use acid flux in the connection. The connection is then heated with a torch until it is hot enough to melt the solder. At this point, the flux 'sweats' out of the joint and touching a piece of solder to the joint will both melt the solder and pull it into the joint. Crimp connectors and compression fittings are also sometimes used in plumbing applications, but soldering is the most common method for houses. It is resists corrosion and lends itself to many different applications.

Soft copper is supplied in rolls. The two advantages of soft copper over the rigid copper are that it is easily bendable to get around corners and obstacles and up through walls, and that it comes in longer lengths which require fewer fittings. It can be flared for connector fittings, unlike rigid copper pipe, and can also be soldered. It is commonly used in attaching water fixtures, faucets in the kitchen and bathroom, and also hooking up dishwashers and ice makers. It is sometimes used in refrigerant lines and HVAC applications.

Note on refrigerant and HVAC usage:

The EPA requires anyone installing and servicing HVAC equipment to pass a certification exam. The test and the training to take it are fairly expensive, and the failure rate is relatively high, so doing this yourself might not be practical.

Rigid PVC pipe that conforms to standards for both pressure and non-pressure applications is available but cannot be used for hot water. It is used for interior water lines, water mains, sewers, drain lines, and vent lines. It is very durable and is resistant to many ordinary chemicals such as acids, bases, and salts. It comes in 10-ft. lengths and, like rigid copper, cannot be bent to go around corners. A wide variety of fittings such as ells, tees, caps, reducers, and threaded pipe connections is available. The pieces are joined with solvent cement which is easy to use. PVC pipe is much less expensive than copper and is used for many water supply installations

ABS pipe is easier and less expensive to install than metal piping. It features superior flow due to its smooth interior finish, and does not rot, rust, corrode, or collect waste. It can also withstand earth loads. It resists mechanical damage, even at low temperatures, and performs at an operational temperature range of -40°F to 180°F. ABS has many of the same physical attributes as PVC. It is used primarily in venting, waste removal, and drains. While it is available with a pressure for industrial use, it is used only in non-pressure applications for houses. It is generally black.

PEX piping is not without controversy, and is relatively new in the US. It has been used extensively in Europe for over 35 years, but in the US many question its uses in certain applications. PEX is made from high density polyethylene. Its uses are in hydronic radiant heating systems (where hot water is circulated through pipes to a baseboard, radiator or in-floor tubes), domestic water piping, and insulation for high-tension electrical cables. One of its major advantages is its flexibility, making it suitable for difficult applications. PEX is often installed using a distribution manifold which connects to the main water supply. This makes it very easy to work on specific areas of the home. For example, you could turn off the water at the distribution manifold to the kitchen sink and make repairs without shutting off the water supply to the rest of the home. If used in a hydronic radiant heating system along with ferrous materials, it must contain an oxygen barrier to avoid rusting of the ferrous materials. This is accomplished by sandwiching aluminum tubing with the PEX. It typically has five layers. PEX uses specially designed fittings and requires a special crimping tool. PEX is now available in the hardware superstores, and many hardware stores and rental shops are now renting the crimpers. The materials are priced lower than copper and the installation time is reported to be half that of copper.

Galvanized steel pipe is generally not used in plumbing houses any longer. It is typically found in older homes and therefore may crop up in a remodeling job. It is no longer used due to its tendency to clog with mineral deposits and scale, causing high pressure drops in the line and resulting in low water pressure. It is also subject to corrosion after the surface has been scratched, exposing the steel to the elements. It was regularly used for buried water supply lines to the house. It is still sold in the standard 21-ft. lengths. It is cut to length using a heavy duty tubing cutter or hacksaw and then threaded for the application. Joints are all screwed, with pipe thread compound to insure a tight water seal. Standard fittings are available, such as tees and elbows. Galvanized pipe is now often used for railings or replacement of existing galvanized pipe in the home. Installation is labor intensive, and due to the difficulty in getting proper fits with making each screwed joint, best left to an experienced pipe fitter.

CPVC Pipe is suitable for hot and cold water distribution and has a 400 psi pressure rating at room temperature and a 100 psi pressure rating at 180°F. It is resistant to many common household chemicals, is corrosion resistant, and is widely accepted by building codes. CPVC pipe and fittings are assembled with a solvent, the same way as PVC, but the solvent for one will not work on the other. CPVC pipe used for drinking water may leave a taste in the water for the first year.

Polyethylene comes in small tubing sizes and in pipe sizes. It is available in rolls of 100 ft. and is pressure rated at 160 psi. It is approved for drinking water use and is therefore another candidate for use as your main supply line, at 20-30 percent of the cost of copper. The tubing sizes are good for ice maker connections.

Braided connectors, while often shunned by the professional plumber, are the answer to the do-it-yourselfers' prayers. These connectors are braided PVC tubing reinforced with an internal woven fabric or external stainless steel mesh. They come with threaded connectors already attached on the ends in sizes suitable for their specific uses. These are used to connect faucets and toilets to the shut-off valve on the water supply. They come in a number of lengths and are close to foolproof. (Nothing to do with plumbing is absolutely foolproof.) For the do-it-yourselfer who does not want to invest in flaring tools or swage fittings, they are a great option.

Choosing pipe to use for any application is important. The use of specific materials is dependent on local building codes. Check with local building officials and inspectors before proceeding in any application. Using the proper pipe for each application makes the job easier, and checking with local authorities makes the application up to code.

Fixtures

One area where you will be tempted to violate your construction budget is the selection of plumbing fixtures. There is a wealth of styles, colors, features, and prices to appeal to even the most determined building materials shopper. In some cases it makes sense to go for better quality, and in others you can safely settle for less.

Tubs and showers — How many modern houses have you seen which have large Jacuzzis in the master bath suite that are covered with dust? These always sound like such a good idea, a giant bubbling pool of soothing hot water with water jets massaging your tired muscles. However, when you have figured the increase of your water and electric bill, it does not sound quite so inviting. Besides, who has the time to relax in one? Before you pick one of these, ask people who have one how much they use it. If you still must have one, consider one the size of a regular tub. It will cost less to buy, install, and operate.

When you are just trying to pick out a bathtub, decide what size and features you want. Do you want cast iron or fiberglass? While this is somewhat a personal decision, an enameled tub is more durable over time. If you choose fiberglass, be sure to check the stability of the bottom of the tub. Step inside one. If you feel the bottom give when you step around, step out and walk away. If it is flexing now, it will continue to do so until you have a crack and possibly a leak in the bottom. While fiberglass can be repaired, a flexing repaired tub or shower will break again. If you are remodeling and putting in a fiberglass shower, look for the remodel styles and sizes. These are designed to fit through the doorways and into the existing bath. Remember that once this tub or shower is in, it is probably there for the life of the house. Do not buy less than you want with the thought of going back later and replacing it. It will cost too much to do that. Buy what you want now and save the money elsewhere.

Vanities and sinks — While the vanity is not technically a part of the plumbing, it holds part of the plumbing (the sink), and the plumbing helps to hold it in place. There are almost as many choices as there are companies that make them. Vanity tops come as pre-rolled plastic laminate, marble, granite, tile, and a host of other materials. A sinks can be one-piece construction with the counter top, or a drop-in. When considering the material, weigh the initial cost against durability. This is especially true in the kitchen, where the counter top gets a lot of wear. Pay particular attention to the edge of the counter top and realize how many times a day something will be dragged across it. While not as difficult to replace as a tub, a counter top, especially one with a molded-in sink, is a major expense. Try to get the best quality you can, while meeting your aesthetic design requirement and staying in your budget.

Toilets — All toilets are now required to be low-flow designs, so you are primarily going to be choosing based on style and color. The cost range variation is not excessive across the basic product lines, but some very high-end products are available. They all adequately perform their required function.

Faucets — This is one area where you can let your artistic side get the better of you. Faucet sets for the bath can run from under $100 to over $1,000. They all do the same thing: start and stop the flow of water into the sink. Granted, some do it with a lot more class and style. In the bath you have an extra choice to make: standard or wide set. Typically, the wide set faucets are more decorative and more expensive. With faucets, you can think about getting a cheaper one now and replacing later. Swapping out a faucet set is a 2-hour job for a do-it-yourselfer. A lot of the cost differences between the top and bottom lines are due to design shape and the finish applied. The basic function of all faucets is relatively the same.

Water Heating — Water heaters have already been discussed in the plumbing savings section. The average superstore has a good selection of water heaters in various styles, sizes, features, and efficiencies. Select the size based on your family size and how you use the water. If you are building a large house for a small family, base the size of the water heater on the number of people who could live in the house, just for resale. Then look at the warranty, energy usage, and features. Manufacturers are now putting a lot of controls on heaters, and these features determine the price range. Do not pay for more than you want or need.

Plumbing Checklist

☐	Review plumbing layout plan to make sure hot and cold water are available in all locations, inside and outside.
☐	Determine what materials you will use for all plumbing applications – hot, cold, pressure, non-pressure, supply, and sewer.
☐	Decide whether you will connect to a public water supply or well, then connect and provide temporary water to the site.
☐	Install rough-in piping including placing large one-piece showers and large tubs or Jacuzzis.
☐	Make sure all vent pipe trees vent through the roof.
☐	Make sure all sewer waste lines slope toward the outside of the house and are properly supported.
☐	Install a PRV on the supply line if required.
☐	Choose fixtures for the bathrooms, and the water heater.
☐	Have HVAC piping installed through framing as required.
☐	Have a rough-in inspection before walls are finished and cover the work.
☐	Do the final trim out after walls are finished and cabinets are installed.
☐	Have a final plumbing inspection if required.

Electrical

> **CAUTION:**
>
> While it may look like wiring a house is simple, it is not. This work should, and in many cases must, be performed by an experienced and licensed electrician. There is more to electrical wiring than meets the eye. Configuring circuits, proper wiring to common and hot sides of the panel, grounding, and GFCI installation are all critical. One wrong wire can lead to fire or death.

Planning

Planning for the electrical system, switch and receptacle locations, lighting, and appliances is critical. Proper planning up front can make living in your house more enjoyable for years to come. Before calling the electrician to come start the wiring, consider the layout and needs for your home.

Your electrical contractor is responsible for all wiring beyond the meter box. The power company is responsible for providing power to the meter box. In some areas the power company is responsible for installation of the meter, but in others, it may be done by an approved, licensed electrician. This is not be done until all electrical inspection has been completed and approved. It is important that no electrical wiring be covered or enclosed prior to the inspection. Electrical inspection is normally done twice, once after rough-in and again before the power is connected. For any new construction, local building codes are very strict about wire types and sizes, breaker types and sizes, junction box sizes, receptacle box sizes, Ground Fault Circuit Interrupter (GFCI) locations, and many other components. Your contractor should have a good working knowledge of the local codes and know all of the requirements. Work closely with your electrician to make sure all your needs are met and within the code requirements.

The rough-in electrical work starts after the rough-in plumbing and interior piping have been completed, and the HVAC units and ductwork are in place. This order ensures that wiring will not have to be moved to allow for plumbing and ducting, which are not flexible. Rough-in includes locating the main breaker panel; installing a box for every receptacle, light, appliance, and switch; then running the wire between them. No connections are made at this time. The wires are all left hanging out of the boxes. The finish electrical work will involve wiring in all switches, receptacles, lights; and connection for the dishwasher, central vacuum, built-in appliances, fans and hoods, and HVAC.

Main service panels (load centers)

In the initial planning for the house, you will have made several decisions which will impact the electrical system. Are you going to have overhead or underground power supply? Will you have gas, oil, solar, coal, wood, heat pump, or electric heat with air conditioning? Are you planning a heated swimming pool, hot tub, or maybe a workshop with power tools?

All of these will impact the amount of power which will need to be supplied to your house and therefore the size of the "service," which includes the wiring size coming from the power company and the size of your power panel(s). The main service panel (sometimes referred to as the breaker box or load center) is the distribution center for electric power to your house. It should be located very near the point at which the power comes in to the house. The wires from the meters come through a conduit and attach to three large metal bars called busses. Circuit breakers clip onto the bus and the wiring is connected to the bus and the breakers. Building codes will govern the minimum service you must have (rated in amps) but you may need more depending on your answers to the questions above.

The most common size panel for new construction is 200 amps. You might need more than one. They come in various configurations and will accommodate breakers of various ratings. Circuits feed out of these panels for both 120 V and 240 V applications. The electrician determines which loads go on which circuits so as to not exceed the breaker rating. If you have special requests, the electrician should be able to work with you as long as it makes sense electrically. Grounding for your electrical system is made at the panel. A heavy copper wire is run from the ground bus in the panel to a copper rod driven into the ground. Years ago, the water lines were used as a primary or secondary grounding system. This worked well when the pipes were all copper or iron, and every joint had metal-to-metal contact. It is not an option with the non-conducting products used for water lines today.

Part of the electrician's job is labeling all the breakers in the panel, so you know which breaker controls which circuit. These lists are sometimes incomplete. Taking the time to make a complete list can save you a lot of aggravation in the future (and probably in the dark).

Conductors

The minimum size (gage) of the wire in your house is determined by the building code. You might, however, want to use a larger wire size in some instances. Remember that the numbers seem backwards; 10 gage has a larger diameter conductor than 18 gage. Why would you want to do more than is required by building code? Have you ever noticed that when the hair dryer is started the lights dim? This is because the surge in power it requires exceeds that which the wire can carry at the available voltage, and the voltage drops slightly until everything is in balance. It is like having your own personal brown-out. It is the electrical equivalent to someone flushing the toilet when you are in the shower.

The longer the run of wire from the panel, the more the total resistance and the higher the chance of voltage drop. The fix for this is to use larger gage wire (smaller number) for the long runs. As with all electrical decisions, the time to do it is now. The difference in the cost of wire size now is small compared to trying to change it out later. Virtually all wiring used today has three conductors which allow all fixtures and switches to be grounded through the ground rod, and also makes possible the use of GFCIs.

Switches

"Where's the light switch?" If you have ever asked this question, someone did not do their job. The building code will regulate some parameters such as distance from the floor, but you can decide how many switches you have and their location. Review the electrical layout drawing carefully. Think about entering and exiting the room, and make sure there is a light switch next to the door, just inside the room. Remember that the boxes for switches will be installed when the walls are only framed. At that point, it might not be obvious which side of the doors will be hinged. A switch behind the door is not the optimum location, but a lot of switches end up there. Consider where you will want three-way switches (hallways, rooms with two doors) and dimmer switches. You may want some wall receptacles to be switched so that the switch will turn lamps off and on. Switches for exterior lights are often overlooked. You might want a switch in the bedroom to allow you to turn on the outside floodlights without walking to the other end of the house in the dark. Would you like to be able to control the bedroom lights from the bed? Now is the time to plan for that. Time spent in studying and determining switch placement will be well spent.

You can choose from several types of switches. The standard is the toggle switch which comes in white, brown, or ivory with matching cover plates. You can get switches that click and silent switches that do not. Also available are rocker switches and a number of decorative styles. Switches can also be lighted — not a bad idea for a bath at night. Dimmer switches are also available in a variety of colors and styles, including several remote control styles. Let your budget be your guide.

Receptacles

While it may be possible to have too many receptacles, it is highly unlikely. Code requires a minimum of one receptacle on each wall and their placement at specified intervals. You might want more. In the kitchen, think about the small appliances you use daily and in cooking. It is also a good idea to have receptacles on several different circuits in the kitchen. For instance, put one on the circuit with the microwave, one on the refrigerator circuit and one with the dishwasher. This will allow you to operate several appliances at the same time without worrying about tripping a breaker.

Plan where you entertainment center will be, along with all its associated equipment, and place extra receptacles there. If you are going to have a shop or workspace in the garage, you will need to have enough receptacles to run all your tools. You might want to have several separate circuits there too. Do not forget your electrical needs outside the house for everything from hedge trimmers to Christmas lights. Weather proof receptacles and switches are required for outdoor use.

All outdoor, kitchen, bathroom and garage receptacles are required to be on Ground Fault Circuit Interrupters (GFCI). These devices interrupt the flow of electricity (trip) when a current to ground is detected. Their intended function is to prevent electrocution. They are required by the building code. There are two types. You can get a circuit breaker which has the GFCI built in. Once installed in the panel, it will protect every receptacle fed from that breaker. The other type is the GFCI receptacle. The receptacle can be wired to provide protection for other receptacles as well. All GFCI devices have test buttons which, when pushed, create a ground fault and trip the breaker. They should be tested frequently. If they fail to trip, they should be replaced immediately. They should be replaced by an experienced electrician, because if they are not wired correctly they will not work as intended, leaving you unprotected. All GFCI receptacles are required by code to be marked.

Miscellaneous Wiring

Several other wiring options should be considered at this time. If you are going to have a security system, have the security company do their preliminary wiring with the other electrical rough-in. This is typically included in the estimate from the security company. Other wiring which the electrician will do for you includes cable and Internet outlets, telephone, any speaker systems or intercoms which you want throughout the house, and home theater or surround sound wiring. When laying out these systems, think ahead to possible future needs. An additional cable connection will cost very little if done now. Do not forget doorbells, the HVAC thermostat, fire alarm systems, and central vacuum outlets.

Lighting

Lighting is another budget breaker. While it is certainly possible to buy a ceiling fixture for under $10, it may be impossible to find one you like for that price, especially when there is a host of more decorative and aesthetically pleasing fixtures. Ceiling light (non-recessed) replacement is another easy do-it-yourself project, so you could go low cost for now, with the idea of replacement in a few years, and save your lighting budget for the major light fixtures like chandeliers for the dining room and foyer. While more ordinary lighting fixtures are relatively inexpensive, many of them are required for a whole house, and the cost can add up quickly. You should have looked at lighting fixtures while you were creating your budget, and budgeted accordingly.

Heating, Ventilation, and Air Conditioning

During your initial planning for the house, you gathered data and cost information to help you determine what options you want for your HVAC system. Determine the most economical and satisfactory solution based on your preferences and locale.

Your decision is based on the climate in which you live, and on fuel and energy cost comparisons are for your area. The most fundamental decision is whether you want heat and cooling. In some climates you might not need cooling. In some you might have only a minimal need for heating. In most cases, you will need at least some degree of both. You can choose to have entirely separate systems, a system that shares some components, such as the fan and ductwork, or a completely integrated system like a heat pump.

Much of your choice will depend on energy availability and cost in your area. Which is more economical — natural gas, propane, fuel oil, electricity, or some other fuel? You might want to use wood stoves or fireplaces for heat or even solar heating. Talk with a heating contractor and/or your local gas, oil, and electric companies. Many times they offer consultation without charge to determine the capacity of the system you need, its installed cost, and an estimate of the operating cost. Often gas and electric companies offer a discount or rebate on the systems and appliances, or low cost financing if you commit to their energy source for your HVAC and water heating. Check the energy ratings on the various systems. For two story houses or houses over 2500 sq. ft., consider a zoned system that incorporates two or more smaller systems instead of one large one. Investigate what system is being used in most new and remodeling construction in your area. If you are used to having gas heat, and are considering a heat pump, remember that the heat pump circulates heated air that is only 90-95 degrees F. While it is an economical system, it might not feel comfortable to you when you are standing over a floor register.

The HVAC system will probably be the largest single load on your electrical system. The ductwork is normally installed prior to the electrical rough-in. Virtually all systems require 240 V service, either for the heating coils, auxiliary/emergency heat (heat pump), or the compressor (heat pump or air conditioner). All systems that utilize refrigerant must be installed by an EPA-certified refrigerant technician who has passed an extensive exam.

After installation, the HVAC system must be evacuated and held under vacuum for a specified time to ensure there are no leaks before the system is charged with refrigerant. Any future servicing of the refrigerant system must be performed by a certified technician.

Electrical Checklist

☐	Review electrical plan drawings with special attention to: • Switch locations • Receptacle locations • High load areas • Major appliances
☐	Determine HVAC requirements and select the system to be used. Have units set in place and ducting run after the house is framed.
☐	Check total load to house to determine the size of the load centers. Allow for non-typical or high load areas for tools and appliances.
☐	Install rough-in wiring after plumbing and HVAC ducts are installed.
☐	Check the rough-in wiring to make sure all the requirements of the layout plan are met.
☐	Make sure all wiring is installed for security systems, intercoms, doorbells, and entertainment systems.
☐	Have rough-in wiring inspected and correct any deficiencies.
☐	Select all lighting, switch, and receptacle types.
☐	After all wall finishing is complete, install switches, receptacles, lighting, and make all final connections.
☐	Check to make sure all breakers and circuits are clearly labeled in the load centers.
☐	Have the final inspection done and the meter installed.

Chapter 11 Checklist

☐	Select the highest quality kiln-dried framing lumber you can afford.
☐	Attach sill plate to foundation.
☐	Attach floor joists to the sill plate and reinforce for heavy load areas.
☐	Install subflooring on floor joists.
☐	Determine material (wood, metal) and technique (stick built, platform, etc.) to be used for wall framing. Construct exterior and then interior walls.
☐	Install sheathing.
☐	Construct roof framing and apply sheathing and roofing. Make sure the roof has proper ventilation.
☐	Check that proper flashing is used in all areas where something protrudes through the roof, and in all valleys on the roof.
☐	Install guttering and downspouts.
☐	Install exterior wall covering – siding, brick, stone, etc.
☐	Complete exterior finishing checklist.
☐	Complete plumbing checklist.
☐	Complete electrical checklist.

CHAPTER 12

Kitchens

For most homeowners, the kitchen is much more than a place to prepare and serve meals. It is the hub of family communication. Many people "live" in their kitchens. You might want a place in the kitchen for your children to study, as well as a desk where you can pay the bills, read the morning paper, and look up recipes on a laptop. You might want the kitchen to contain both a bar for quick meals, and a seating area for serving more elaborate meals.

The style of a kitchen commonly reflects its architectural nature. For example, if you are building a blocky Greek Revival home, your kitchen cabinets might have symmetrical panel doors of varying sizes. For an art deco theme, you will probably choose clean lines with glass doors or metal accents.

Aside from the architectural style, there are many considerations when designing a kitchen.

Size: The kitchen should be large enough to provide sufficient counter and work space. Every cook has different requirements. People who do a lot of baking need ample counter space; those who regularly cook gourmet meals need extra storage for their special tools. Those who cook very little will not want to waste square footage on the kitchen. Your kitchen should accommodate the features and appliances you want and have enough room to move around, without extra steps or distance between major workstations.

Purpose: Besides cooking, the kitchen will probably be used for at least some of your entertainment, as well as for eating. Are you comfortable with only a small snack bar, or do you want space for a breakfast nook? Do you need a full-sized table in the kitchen? Will there be multiple range hoods, or perhaps an island? These decisions will influence the size of the kitchen. If your family is used to congregating in the kitchen, they will feel cramped in a small kitchen. Think about how your kitchen will be used ahead of time.

Decisions about the kitchen include:

Cabinets: Do you want base, wall, or corner cabinets? Will the corner cabinets include lazy susans? Do you prefer open cabinets? Glazed or color? Special features?

Appliances: Appliances take up more space than anything else in the kitchen. Because of this, most homeowners select them first to determine space requirements for the layout. Determine which counter appliances you want to use (mixer, coffeemaker, microwave, blender, etc.) and where they will go.

Counters: The type of countertop you choose will affect the look and feel of your kitchen. Many choices are available. Consider:

- Price
- Weight
- Durability
- Upkeep

To save on costs, many homeowners customize each work zone with the type of surface that will be appropriate for use, and put luxury countertops only where they will have an aesthetic impact.

Some of the more common choices are:

Laminate: Laminate is easy to install, and is available in many colors and textures. It is stain and impact resistant. It is also the least expensive countertop, starting at $10 per sq. ft. On the negative side, laminate has visible seams that you will not be able to conceal. If it becomes scratched or marred, there is almost no way to repair it. Also, you cannot cut with a knife on the countertop if your counters are laminate.

Butcher block: Made of rock maple, butcher block costs about $16 per sq. ft. The surface is great for cutting, and as you use it, it develops a unique character from the marring. It can be almost instantly brought back to new condition by sanding it down a little and adding a coat of mineral oil or beeswax. A butcher block countertop is a good choice, but it is vulnerable to water damage. It must be wiped dry, and when used for cutting raw meat or fish, it must be cleaned thoroughly.

Ceramic tile: Ceramic tile is a popular choice because it is easy to install and repair. It comes in almost any design and resists moisture, scratching, heat, and stains. However, the grout used between tiles can become stained or mildewed, and is difficult to clean. Newer epoxy grouts are available which will not stain or mildew. Although they are slightly more expensive, they are well worth the extra cost. Ceramic tile can cost up to $80 per sq. ft. — so the budget-conscious homeowner should be careful about choosing tile.

Engineered stone or granite: These are the toughest and least porous materials. They are resistant to scratches, stains, heat, and the engineered stone will never need sealing. The cost will be at least $50 per sq. ft. and can run well into the hundreds. These countertops require professional installation, and if damaged, will need professional repair. They are also extremely heavy, and this must be taken into consideration if the kitchen is not placed on a slab.

Marble: Marble offers a cool, non-stick surface that is perfect for the serious baker. The cost is fairly low, around $50 to $75 per sq. ft. However, marble is porous. That means it is easily stained and scratched, and may become discolored. It must also be professionally installed and sealed on a regular basis.

Solid surface acrylic: Solid surface acrylic countertops are non-porous. That means there will be no germs living anywhere in the counter, and it is difficult to stain. Clean-up is easy, and scratches can easily be buffed out. It comes in dozens of colors and patterns, including some resembling stone. The seams blend easily with edges and integral sinks. However, solid surfacing can be expensive, especially since you will want to choose one of the best brands to get a good quality counter. It requires professional installation, and many homeowners find that solid surfacing is damaged by hot pots or pans.

Concrete: Some contractors who want to create an unusual shape to their countertop prefer concrete. It can be cast right in the kitchen, is heat resistant, and can be tinted nearly any color. Newer types of concrete have less propensity to crack and offer decorative finishing. However, the price of a concrete counter is high,, unless you are adept at pouring it yourself. If done incorrectly, it can crack, and the look can turn out to be somewhat industrial if it is not planned carefully.

Storage: Do you want your pantry space in or near the kitchen, or would you rather it be a part of your cabinets? Remember, you will also be storing things like pots and pans, canned goods, small appliances, serving dishes, china, flatware, glasses, towels, containers, and cooking utensils.

Remodeling a Kitchen

Every kitchen design should be planned around the classic kitchen triangle, which describes the way you flow from one area to another during kitchen work. The points of the triangle are:

- Sink
- Stove and oven
- Refrigerator

You should be able to reach each of these three points easily, within one or two steps, and there should be nothing blocking your movements from one to the next.

When demolishing an old kitchen, remember that lead-based paint or asbestos might be present. These will need special handling, and may require special waste stations because not all trash yards accept hazardous materials.

If you are remodeling the kitchen, check with the designer and the contractor to see who is going to obtain the necessary permits for electrical, mechanical, or building work. This must be done prior to the beginning of construction.

Plan the Design

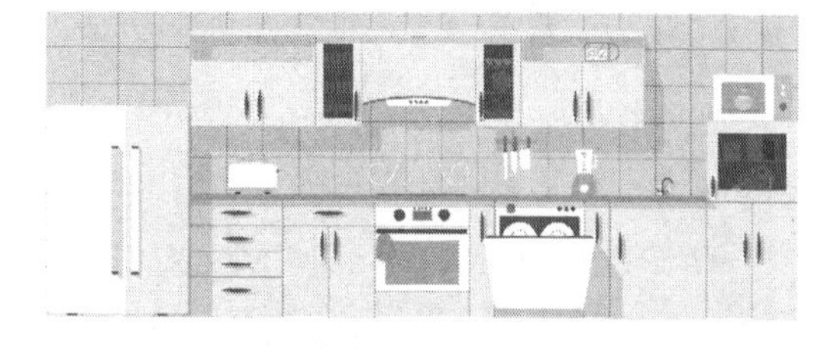

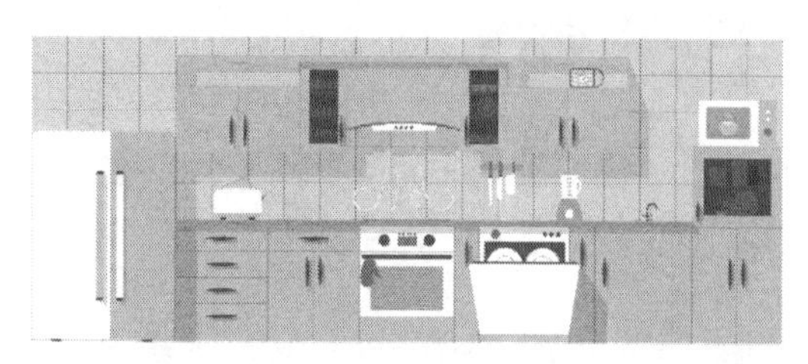

Even though the kitchen probably has a fairly established design, consider changing it. Kitchen remodels hold their value, and if you are planning to sell the house, they are one of the best choices for updating. Designing a kitchen is fun. There are only five basic design plans, and if you are remodeling you will probably already have an idea of which one to use.

The One-Wall Design

If your kitchen is long and narrow, it may have been built in a linear fashion, with the sink, stove, and refrigerator on the same wall. A one-wall design that allows for unimpeded traffic flow, but if there is enough square footage, consider moving one point of the triangle to the opposite wall.

The Corridor Design

Mostly seen in apartment living, the corridor plan has a triangle, but is somewhat crowded. The refrigerator is at one end, with the range and sink directly across from each other — and very little room between. If your kitchen space is at a premium, this is the most affordable way to design the kitchen.

The L Shape

An L-shaped kitchen arrangement features a refrigerator on one end of the L and the range on the other, with the sink located near the center on one leg or the other. It gives more counter and cabinet space, but the corner counter may be hard to reach.

Double L

A double-L layout contains a small workstation used for food prep, often with a cooktop and a second sink, within the larger work area. There plenty of open counter spaces, though no more cabinets than the typical L-shaped kitchen.

The U Shape

A U-shaped design gives the cook good work flow and plenty of cabinet space. Designed like the corridor, but with the end closed, there is room for the range or even a sink on the short end.

The kitchen remodel might require that you add joists to strengthen the floor to support heavy appliances, granite countertops, or a kitchen island. The new kitchen might have new features like an automatic ice-maker or an island with a sink, or you may have chosen to move the main sink. If so, you might need to rough in new water supply lines and drainage pipes. If you are doing this yourself, it may be slow going. Consider hiring a professional plumber.

The electrical service should be at least a 200 amp to support the kitchen. An electrician will probably want to run new wires for lighting and appliances. Code requires that the wiring meet current standards. Be sure to have an inspector approve the electrical system before hanging drywall.

You will need to store your new materials somewhere until it is time to install them. Find out whether the designer or a sub has a warehouse where products are normally delivered. Instead of sending old cabinets and used fixtures to the local landfill, consider using them as extras in the laundry or garage area. You can also donate such products to local salvage centers, or to non-profits that use them for rehab projects.

Use clean, sturdy boxes to pack up the items that will not be used during renovation. Store these away from your project to protect them from damage. This will include pictures, mirrors, furniture, and other items that are in the kitchen or in adjacent rooms. The vibration from the construction can damage them if they are not removed. Before renovation begins, pack up the contents of your kitchen. Start with the items you use the least.

When the day for demolition arrives, pack the most frequently used items in the most accessible containers. Dust is one of the biggest problems when revamping a kitchen. Cover the floors in your home and hang a dust curtain over doorways. Let workmen know which ways you want them to enter and exit the workspace, and be sure that they always use these paths. You can ensure this by taping plastic sheeting over other doorways. Either tape over the HVAC vents, or do not let the central heat run while the workers are sanding; it will blow the dust throughout the ductwork and you will have dust forever.

Chapter 12 Checklist

☐	Budget
☐	Design and Layout
☐	Demolition (if remodel)
☐	Rough framing
☐	Electrical
☐	Number of circuits and electrical load
☐	Major appliances
☐	Wall outlets
☐	GFCI for all outlets
☐	Lighting • Main • Task • Accent
☐	Gas service
☐	Plumbing
☐	Sink
☐	Faucets
☐	Dishwasher
☐	Icemaker
☐	Filters
☐	Heating and air conditioning
☐	Fans and ventilation
☐	Insulation
☐	Drywall
☐	Windows
☐	Doors
☐	Finish/stain

☐	Cabinet selection • Base • Wall • Tower/storage • Open shelving
☐	Island base and overhead
☐	Valances
☐	Crown moldings
☐	Walls
☐	Baseboards
☐	Countertops
☐	Sink
☐	Backsplash/tile
☐	Appliances • Garbage Disposal • Microwave • Refrigerator • Cooktop • Oven • Freezer • Dishwasher • Other
☐	Flooring • Structure • Covering
☐	Paint
☐	Walls
☐	Woodwork
☐	Wall covering

CHAPTER 13

Bathrooms

Bathroom Sizes

When looking through glossy home magazines, who does not pause and dream when looking at a centerfold photo of a luxurious master bath? In the magazines, these rooms look very desirable, but if you stop to calculate the square footage they consume, and the cost of filling that square footage with tile and fixtures, you will quickly reconsider. Unless you are captivated with the idea of bathing in a gymnasium, you can probably make do with a much smaller space.

If all the houses in your neighborhood have huge bathrooms with sumptuous fittings, however, you probably want to remodel yours in this manner. In some areas, a large master bathroom is practically a requirement. In others, it is a luxury.

You might be remodeling a master bathroom, trying to make use of the available space, or for new construction, trying to make the best use of the allotted space. Instead of laying out a hangar-sized bathroom and outfitting it with every available gadget, you might be wiser to utilize some of the space to increase closet size (maybe with a walk-in closet), adding storage, or increasing the usable size of the bedroom. This could also give you more money to fill the space you do use for the bathroom with better tile, fixtures, and flooring.

A toilet, sink, and shower can fit into a space as small as 5 feet x 8 feet. If you need a double sink and a shower/tub combination, the minimum will stretch to about 8 feet x 9 feet. Code requirements define the minimum amount of space for each fixture in the bathroom but adding just a few inches here and there can make the difference between feeling cramped or comfortable. Most codes require 21 inches in front of sinks and toilets, but 24 inches is much better and 30 inches will feel very comfortable. Remember that these clear spaces can overlap, as you are not likely to be using the toilet and the sink at the same time. Having more than 36 inches around a fixture is a waste of space which can be put to better use elsewhere.

To figure out how much space to allocate for the bathroom, sit down with your partner and answer a few questions. Do you have similar schedules, or does one of you go to bed earlier and get up earlier? A dressing area lets you turn on the lights, get dressed or undressed, and find clothes without disturbing the other. Do you both need the sink and the mirror at the same time? A double vanity takes up a lot of room but may save a marriage. If you always shower, you might not need a tub in the master bath, but if you enjoy a good soak, a tub is a requirement. What about a Jacuzzi? How often would you use it? These take up valuable real estate, put a large dent in the budget, and often become merely bathroom ornaments. Your bathroom needs to be sized for your lifestyle and the way you will use it.

Simple or Elaborate

The selection of components for finishing baths is almost unlimited, and prices range from economical to extremely expensive. As you look through the hardware 'superstores', supply house displays, and catalogs, consider usability, practicality, durability, appearance, initial cost, and the ever-important resale value. As with a kitchen remodel, the cost of remodeling a bath can be recoverable or even add value, if you make choices wisely. But just as overbuilding for the neighborhood with a new house or addition can result in your not being able to recover your costs, overspending on bath components can limit your ability to recoup the cost. Remember this as you are preselecting components during the budgeting process.

- **Cabinets and lavatory** — Your cabinet and lavatory layout will be determined in part by the space available after the fixtures are laid out. Cabinets come in many styles and lavatories can range from a freestanding basin or wall-hung sink to a one-piece top on a cabinet (see Countertops). Consider your storage requirements for grooming accessories such as curling irons and hair dryers, toothbrushes, razors, cosmetics and lotions, medicines, towels and paper products, and so on. Having to leave the bathroom to get something you need as you get ready for work every day is impractical. The selection of a cabinet finish is an opportunity to coordinate the colors and textures in the room along with the tile, counter tops, and flooring. Cabinets come in a wide range of prices and, by shopping around, you can find a good look at a good price.

- **Mirrors** — Mirrors are an important component of the bath. They need to be adequate for applying make-up, arranging hair, and other personal grooming tasks. A large mirror reflects light and also give the room a larger feel. Mirrors are available from glass shops as plain plate glass mirrors in any size you want, and offer a low cost solution which has a good look. They are also available in combination with a built-in medicine cabinet, and with side 'wings' which fold out, allowing a view of the back of your head. They come with lighting and without, framed, beveled, or with simple seamed edges. Your choice is determined by your style and budget.

- **Lighting** — I have never heard anyone complain that there was too much light in the bathroom. The bathroom might need to have the best room lighting in the house. Place lighting so that there is ample light to use the mirror without having glare. General lighting should be provided throughout the bathroom, with task lighting at every functional area. Decorative lights often look great in the room but provide inadequate lighting where it is needed. Be sure to check the allowable wattage of the fixtures you choose.

- **Ventilation** — Moisture is always a problem in a bathroom, particularly if there is a shower. Unless provision is made to remove it, there will be problems with mold, mildew, and peeling wallpaper and paint. A ceiling fan which is ducted to the outside is preferred, as it completely removes the humid air from the house. A vent which discharges into the attic can just transfer the problem from the bath to the attic. In cold weather, there will undoubtedly be condensation in the attic, which will cause problems later. Fans are available alone or as a combination fixture which includes a light and/or heater. This is a good opportunity to get additional lighting in the bath at a very low incremental cost.

- Countertop and sink — Countertops and sinks are available in many styles, materials, textures, and colors. Some of the more commonly used are:

    - **Laminate/drop-in:** You can purchase sections of rolled-edge laminate-covered countertops, which can be cut and installed by the average do-it-yourselfer. A hole is cut and a drop-in sink installed (also available in a variety of styles and materials). This is probably the low cost choice. Another style has a squared edge. This style can be built by constructing the plywood base in place, and then gluing the laminate on to the base and trimming the edges with a special router bit.

    - **Stone:** A currently popular material for use in countertops is stone, often marble or granite. For many applications the countertops are made in one piece for a seamless finish, and can be purchased with a shaped edge in several styles. Granite is the most popular choice for counter tops because it is much harder than marble and has greater stain, heat, and water resistance. It is also more resistant to abrasion. The price varies based on the complexity of the edge desired. Drop-in and bottom mount sinks can be used.

- **Engineered stone:** Engineered stone is a quartz-composite product mixed with colored pebbles, polymers, and epoxy. It has an even pattern and more color options than natural stone. It is an extremely durable product that takes the heat and resists stains. Engineered stone is installed using epoxy and costs about the same cost granite. Drop-in and bottom mount sinks can be used.

- **Tile:** Tile comes in a seemingly endless variety of sizes, colors, textures, and patterns. Ceramic tile offers a smooth, glossy, and easy-to-care-for finish, and is probably the most popular choice for walls and backsplashes. It is also used on counters. Clay tiles have a matte finish and are available in many colors and textures. They are equally waterproof and stain resistant. Epoxy grouts which will not stain are now available for use on countertops and are recommended. Drop-in sinks are used with tile.

- **Solid surface:** Solid surface countertops have seamless acrylic faces with seams that are only visible from the underside. With solid surfacing, sink, and counter materials can be integrated to create a fluid, graceful line. Solid surfaces come in a rainbow of colors, patterns, and styles, including stone and glass look-alikes. They are stain and heat resistant, with more edging and border options than natural stone. Scratches are easily softened with a non-abrasive scrubbing pad.

- **Plumbing fixtures** — See the section on Plumbing in Chapter 11.
- **Bathtub/shower** — The provision for bathing can be made in four basic ways:

- **Combination tub/shower unit:** A fiberglass/acrylic combined unit includes the tub and wall sections. These come in one-piece units for new construction, or with separate wall panels that can fit through doorways for remodeling. Most have molded-in soap dishes and handles and are available in different colors. These are easy and relatively inexpensive to install.

- **Bathtub with wall treatment:** These are essentially the same as the combination units, except that instead of having the wall section as a part of the package, the tub is installed and then a wall treatment, such as tile, is used. This gives an expanded choice in tub selection and material (enameled cast iron for instance) and a much greater opportunity to coordinate with tile selections for the tub surround.

- **Separate stand-alone shower and/or tub:** If you have the space and budget, a popular option is to have separate units. You can have a tiled walk-in shower and a separate tub.

- **Jacuzzi tub:** As previously mentioned, a Jacuzzi is a high-cost option, from the initial purchase cost to the installation and operation. Before adding one of these to your bath, be sure you will use it enough to justify the expense.

- **Flooring** — Many choices available for flooring in a bath. The main choices are wood, tile, and carpet.

  - **Wood:** Available in a variety of woods, colors, and finishes, wood flooring is a popular choice for many houses. It is easy to clean but is subject to water damage.

  - **Tile:** Also available in many colors and textures, tile is easily cleaned, but the grout should be sealed to prevent stains from dropped make-up and other accidents. Tile is waterproof and can survive spills.

  - o **Carpet:** Carpet is subject to water damage, mold and mildew. While it is not unusual to see it used in baths, thought should be given to the maintenance requirements.

- **Electrical outlets** — Just as there is never too much light in a bathroom, there are never too many electrical outlets. Number and location might be dictated by the local building code, and circuits must be GFCI protected. Determine how many appliances will be plugged in at once (curling irons, hair dryers, radio, and phone charger) when calculating how many you need. Codes do not allow outlets to be placed within a specified distance of bathtubs and showers.

Tile work

Like the right accessories to match an outfit, the right tile influences the impact of your bathroom. Whether you are using colored ceramic, a clay tile, or a pattern, tile can bring texture and enhance the overall look of the bathroom.

Tile selection

Tile is one of the most popular wall treatments in bathrooms, particularly on surfaces which will be exposed to water. Tile is a durable surface, within the ability of many do-it-yourselfers with a little training. It is relatively easy to keep clean and comes in a myriad of colors, styles, and designs. Before you go to the tile dealer to select your tile, make some preliminary decisions about what you want. You can choose from glazed ceramic tiles, matte finished clay tile, and stone in either plain of textured surfaces. Narrow the scope of what you want before you go into the store. However, no amount of pre-visit planning will prepare you for the thousands of available patterns, colors, and textures at most tile and carpet dealers. You will undoubtedly find many tiles from which to make your final selection. Ask the dealer about ease of cleaning, durability, tendency to spot or discolor, and care requirements before making your final decision. Take samples of wall coverings with you to make sure all the materials blend or match as you expected. Most dealers allow you to check out sample tiles and take them home (for a remodel) so you can see how they match the fixtures in your bathroom.

Design

Tile designs are nearly as limitless as the selection available. You can use plain tile, add a border, intersperse tiles with patterns or designs, alternate colors, and so on. Most tile options add a lot to the interest of the room and not very much to the cost. Tile dealers typically have a lot of literature and suggestions, so take advantage of their experience.

Installation

Many hardware superstores offer free classes on tile installation which teach you how to produce professional-looking results with a few special tools. Tile cutting saws can be rented; the other tools are reasonably inexpensive. If you are doing a remodeling project that involves removing old tile, take safety precautions. The primary means of removing the old tile from the wall is with a hammer. Broken tiles are EXTREMELY sharp, as sharp as razors. Small chips fly around, so good eye protection, like sealed goggles, are essential. This cannot be over-emphasized. Bathtub fixtures and the tub must be protected from damage from falling tile pieces.

For tubs and showers, a cement backing board should be attached to the wall, and the tile installed on the board. This product is waterproof, will not swell and crack, and provides a stable surface with good adherence for the tile. It is attached to the stud wall with galvanized screws.

Careful layout of the tiles is crucial for a do-it-yourself job. If you end up using partial tiles to finish a row (almost a certainty), you want equally sized pieces on each end. You must therefore start in the middle. This is only one of the things you will learn in the tiling class. Instruction is also available on video and several websites, but there is no substitute for hands-on work.

Chapter 13 Checklist

☐	**FIXTURES**
☐	**BATHTUB/SHOWER** • Large enough to accommodate each user • Non-slip bottom surface or floor • Faucets within comfortable reach and properly oriented (hot on left) • Soap, shampoo and conditioner storage, and towel bar conveniently located • Shower head at a comfortable height for all users • Easy and safe entry and exit • Grab bar in the tub or shower • Sufficient clear space in front of tub/shower • Glass shower doors with safety glazing • Shower door that swings into bathroom, not into shower
☐	**TOILET** • Located a minimum of 16 in. from center line to wall or cabinet • Sufficient clear floor space in front • Doors do not open into toilet space • Paper holder comfortably located • Toilet design allows ease of cleaning
☐	**LAVATORY** • Sink height comfortable for all users • Distance from center of sink to wall is a minimum of 15 in. • Minimum of 30 in. clear space in front • Door does not swing toward cabinet and block access • Backsplash or tile is waterproof and sealed to lavatory

☐	**WALLS** • Walls that come in direct contact with water have a waterproof covering (tile, fiberglass, stone) which is sealed to prevent unseen leaks and for ease of cleaning. This includes enclosures around tubs, showers, and behind sinks. • Walls which are subject to splashing have water resistant covering (wallpaper or enamel paint).
☐	**FLOORS** • Finished in a way that resists water damage • Easy to clean • Non-slip surface
☐	**MECHANICAL/ELECTRICAL** • An efficient ventilation system • Bathroom will be comfortably warm? • Sufficient towel bars, clothes hooks • Sufficient electrical outlets, all with GFCI protection, located away from the tub/shower • Adequate lighting for all bathroom activities • Shutoff valves on all water lines to sink and toilet
☐	**STORAGE** • Adequate shelf or counter space around the lavatory • Sufficient space to store toiletries, grooming equipment • A space for towel storage in or near the bathroom • A place to store medicines and cleaning supplies out of the reach of small children

CHAPTER 14

Putting on the Finishing Touches

Drywall

Installing drywall, while somewhat strenuous, is not a difficult job. So long as you have a helper for lifting the drywall into place, the actual attachment is relatively easy. Finishing the drywall will take a little practice to become proficient.

Purchasing Drywall

Determining the number of sheets of drywall you need is relatively easy. Just determine the number of square feet of wall and ceiling you are going to cover, and buy that much drywall, allowing for scrap. You can consult the dealer for help in determining number of the rolls of tape, buckets of drywall compound, and nails or screws you will need. If you have outside corners, each one will require a metal cornerbead.

Hanging Drywall

Most drywall comes in sheets that are 4 ft. wide and 8 ft. long, but is also available in 10 ft. and 12 ft. lengths. It is typically oriented with the long dimension vertical (on the walls). However, the objective is to minimize the length of joints to finish, so it is perfectly acceptable to hang it horizontally if that accomplishes the goal. Drywall of 1/2 in. thickness is usually used for walls, 5/8 in. thickness for ceilings. Carefully lay out your drywall sheets to minimize joints. Measure and cut the drywall sheets with a sharp utility knife. After you have scored it on one side, lay it on a piece of 2X4 and snap it along the cut. Then turn it over and trim the paper on the other side.

Begin by installing the drywall on the ceiling. Before starting on the ceiling you will need to construct a pair of "T" braces. You can make these with a 3-ft.-long piece of 2X4 nailed at 90 degrees to the end of a length of 2X4 which is about an inch longer than the distance from the floor to the ceiling. These braces will be used to lift the drywall for the ceiling into place, and wedged to hold the drywall in place while it is being fastened. Before raising the sheets into place, cut out the required openings for lighting, fans, and other necessary items. Fastening is done by either nailing or screwing into the ceiling joists; using screws is the most common practice. If you use a nail, be sure to use one designed for drywall, and hammer it below the surface of the drywall so that the head of the hammer leaves a dimple in the surface. If using screws, the screw heads should also be below the surface. These dimples will be filled during the finishing process. Fasteners should be spaced at 6 in. intervals starting at the center of the sheet. Using a chalk line will help you get the nails or screws in the center of the joists.

Once the ceiling is done, you can start on the walls. Before hanging drywall on the wall, carefully lay out and remove material where windows, doors, switches, and receptacles are located. Once the holes are marked, you can start the hole with a drill and finish it with a keyhole saw. Butt the wall panels tightly against the ceiling. Wall fasteners begin 4 in. from the ceiling and are spaced 6 in. apart. When you have an outside corner measure, cut a metal cornerbead and fasten it in place. You do not need to dimple the fasteners on the cornerbead because they will be covered with plaster.

Joint Taping

The joints where pieces of drywall meet are finished using a process called taping. To have a smooth surface, the joints must be filled and finished so that they at least appear to be flush with the rest of the wall. To facilitate this, the edges of the drywall are manufactured tapered. To reinforce and smooth the joint, a layer of joint compound is first spread into the recess created by the tapered edges using a 5 in. taping knife. Smooth the compound until it is flush with the rest of the sheet. Then center the drywall joint tape over the joint, and press it into the compound. Compound will squeeze out through the holes in the tape and along the edges, so make sure there is still some compound under the tape. Once the tape is firmly in place, smooth out the compound with your taping knife. After you do this, fill all the fastening dimples with compound and smooth with the taping knife. Also apply compound to the cornerbeading, using the 5 in. knife to smooth the compound from the edge of the bead, over the fasteners, and out onto the board. Allow all the compound to dry completely, usually for at least 24 hours. It is important not to apply too thick a coating at one time, since the compound will shrink and crack.

After the first coat of compound dries, you will need to apply two more thin coats. The first should extend a few inches beyond the first coat. After it dries the third coat should extend about 6 in. to each side of the joint and should be done with a 10 in. blade. After the third coat is thoroughly dry, sand and smooth all surfaces lightly with medium grit sandpaper.

The inner corners and the joints between the wall and ceiling must also be taped. Tape is available that is pre-scored down the center and folds easily to facilitate this. The procedure is the same as for wall joints, except that more care must be taken in the corners not to cut the paper with the edge of the joint knife. Sanding is also required after the third application of compound.

WARNING: Sanding generates dust.

If you are doing this as a remodeling project, be forewarned that sanding dried joint compound is akin to tossing up a box of face powder and letting it hit the floor. Dust goes everywhere and gets into everything. Tape plastic over doors leading to anywhere else in the house, and tape over any HVAC ducts. You should also wear a dust mask while doing the sanding and clean-up. An exhaust fan in a window to give the room a slight negative pressure is also a good idea.

Painting and Finishing

Painting is one of the areas where people think they can save money by doing it themselves. While this is probably true, you also have to consider time as having value. A paint crew can come in and paint an entire house, including woodwork, in 2-3 days. The do-it-yourselfer may require a couple of weeks to complete the job. This can seriously delay your schedule and cause you to lose subs to other jobs while they are waiting for you to finish. You have to make this decision for yourself, but consider the real cost of saving the money.

The array of wall finishes from paint to faux finishes, stone to wallpaper, rough cut cedar to fine mahogany, is endless. Your choices are limited only by your sense of design and your budget. The most common wall finish is paint, which is discussed here. Instructions for the application of specialty materials are available where you buy the product. There are few finishes the average do-it-yourselfer is not able to accomplish. Many people feel that wallpaper is beyond the ability of the average person, when, in fact, putting up modern pre-pasted papers requires only a little on-the-job experience, and a good straightedge, paper tray, razor knife, and sponge. In most cases, especially where the wall starts as new drywall, applications begin with a sealer.

For drywall, it is recommended that the compound be allowed to dry for 5 days before painting. The first coat that is put on a new drywall surface is a sealer coat. After the sealer, sand very lightly with fine grit sandpaper.

The next coat is a primer coat. Primer with a mildew repellent included should be considered in damp areas or anywhere mildew or mold could be a problem.

You can have the primer tinted to reduce the number of coats required if you are planning to have a dark-colored paint finish coat. Primer does not take a tint like finish wall paint. It will not be the color you expect, but may be close enough to let one top coat do the job. For instance, my dark chocolate brown tint turned out mauve in the primer.

Again, lightly sand after the primer coat with fine grit sandpaper. Remember to wipe down the walls with a tack cloth after each sanding to remove the dust and particles. If you are papering, the wall is now ready for the paper. If you are painting, the final coat will be wall paint in the color of your choice. Consider the use of the room when selecting paint type. Most often people use flat wall paint with semi-gloss enamel in painted woodwork for just a little sheen. This combination works well in a living room or bedroom where the walls are not subject to a lot of dirt and moisture, and will thus survive with a little sponging now and then. For kitchens and bathrooms, consider using semi-gloss enamel on the walls and possibly ceilings. This is better for resisting moisture, and will withstand some hard scrubbing to remove grease and food splatters.

Ceilings

Ceiling finishes are somewhat more limited than wall coverings. Your choices are probably limited to either paint or one of the modern textured finishes. For a painted ceiling, the surface with all its joints must be carefully taped and finished. Unlike walls, where in the finished room only small sections are seen at one time due to wall hangings, pictures, furniture, windows, and doors breaking the surface, the ceiling is nearly all visible at once. In addition, it is usually seen in a raking light (light shining from the side) which shows every hill and valley in the surface. It is for this reason that the textured ceilings are frequently used. If you do plan to paint, follow the same procedure for the ceiling that you used for the walls.

If you are going to use a textured finish, perfect flatness is less of a concern. There are several textured finishes to choose from; probably the easiest to apply is the spray-on finish. This is a thinned-down joint compound mix which has small, irregular-shaped particles that give the texture its nubby look. There are also special texturing compounds made just for this purpose. This is a fairly easy application for the do-it-yourselfer using rental equipment. Do this type of ceiling before painting the walls, and tape the edge and tape plastic sheeting over the walls to prevent spray from getting on them. Follow all safety instructions that come with the spray gun and material, especially wearing goggles and a breathing mask.

Another texture is made by applying joint compound on the ceiling with a trowel and then pressing a sponge into the material to form a pattern. A light touch with the sponge yields a smoother surface with a more professional and finished look. You can also create a stipple finish or a swirl pattern using a stiff brush on the applied joint compound.

Woodwork and Railings

Woodwork and interior railings can either be stained or painted. Painting can give a more formal finished look and allows you to use a less decorative (i.e., less expensive) wood than staining. Whichever method you choose, save time and frustration by painting or staining the woodwork before attaching it to the wall, then filling the nail holes and touching up with paint.

Garage Doors

There are many styles and finishes of garage doors to choose from. If the garage is not going to be heated, you can save some money by getting uninsulated doors. Garage doors are available in several styles to match your house style. They are available in wood and galvanized steel panels. Most are pre-finished and some come in a variety of colors. They also come with a variety of windows, from none to any number of styles.

The size of your garage determines whether you have a single large double door or two singles (for a two-car garage). While a single large door will save you money, at least on installation, remember that a single double door will probably result in the cars being closer together. Double doors are typically 16 ft. wide. Singles come in 8 ft. and 9 ft. widths. The 9 ft. width will allow more leeway in putting a large car or truck in the garage. Door installation can be done by the do-it-yourselfer, but if you selected one with a large spring mounted on the wall above the door which has to be wound up, have it installed by an expert. Winding the spring is a dangerous operation, and many people are injured each year trying to do it without training.

Pools and Equipment

The installation of a pool is a major financial consideration. Not only is the installation expensive, but upkeep can be costly. In some parts of the country, having a pool is a liability for resale. For many families, a pool is an integral part of their lifestyle and provides immeasurable recreational value. In some areas, pools are considered a necessity.

Pools come in many sizes and styles, including freestanding above-ground pools, in-ground pools with liners, in-ground fiberglass pools, and in-ground concrete pools. Planning for the building of a pool is like planning for any other construction project. You need to check on codes and restrictions, make a budget and schedule, look at materials and designs, select the subs, get the financing, and complete the project. The pool design and installation will probably be done by a contractor who specializes in swimming pools and can recommend the sizes and specifications for pumps, filters, and cleaning equipment. The contractor can also instruct you in water testing, treatment, and maintenance requirements.

Decks

If you are planning a deck, the design and configuration have undoubtedly been done as a part of the design for the new house. Adding a deck is a popular remodeling project, or you might be replacing an existing deck which has deteriorated and is no longer attractive or safe. Once you have the design for the deck, the question is what material to use. The basic choices are natural wood and composites. Natural woods include:

- Cedar
- Redwood
- Mahogany

- South American hardwoods
- Salt treated pine (pressurized lumber)

Availability and cost of these options may vary in different parts of the country, but the salt-treated lumber has become the low-cost standard most places. The choice depends on the look you want, what is available, and cost

Composite deck materials in a variety of colors are relatively new in this market. They have a lot of "green" power since they are made of recycled materials. Though many offer impressive warranties, most composite decks have not been in place long enough to see if they are as truly low-maintenance and durable as the manufacturers claim. In addition, the composite boards are not as strong as natural wood and require more support (support joists placed closer together). Since the supports are natural wood, the true life of the deck is the life of the support wood. Decks are not normally painted, but they can be. Do not use non-pressurized framing material and count on the paint to prevent weather damage. Get advice from your local paint supplier about the best paint for the wood you are using for your deck.

Chapter 14 Checklist

☐	Hang drywall, cutting all required openings.
☐	Tape the joints and finish with joint compound and final sanding.
☐	Seal, prime, and paint drywall
☐	Install wallpaper, paneling, and other wall coverings.
☐	Paint or stain all woodwork and railings, then install and touch up paint.
☐	Select garage doors and install.
☐	Plan for pool design and installation as a separate construction or remodeling project.
☐	Complete design and planning for decks.
☐	Make decision on type of materials to use.
☐	Construct decks.

CHAPTER 15

Potential Hazards and How to Prevent Them

Radon, Asbestos and Lead

These three environmental and health hazards are discussed in Chapter 8 under "Environmental Testing." For new construction, the EPA has recommended guidelines for the prevention of potential radon problems and ease of installing radon control systems. The recommendations vary depending on the zone you are in, relative to the risk of high radon levels indoors. You can see the EPA map at www.epa.gov/radon/zonemap.html. Check on the building codes for your area to see if preventive measures are required. If you are in Zone 1, you might want to take preventive measures, even if codes do not require it, to reduce the cost of eliminating high radon levels later.

Mold

You are much more likely to be faced with mold in the renovation of an old house than the building of a new one. Mold is any of several types of fungi. Molds are present in indoor and outdoor air, and the particles, or spores, are much too small to see with the eye. They grow on any number of materials in the presence of moisture. The materials on which mold grows include carpet, wood, paper, food, and even insulation. The spores are much like seeds and travel through the air looking for a place to grow. If the house you are renovating has had roof leaks, damp basements or crawl spaces, plumbing leaks, inadequate venting in bathrooms, and/or dryers not being vented to the outside, there is a likelihood that mold exists.

What is the problem with mold? One problem is that the material on which it grows may be damaged. A larger problem is the health hazard it presents. The spores are inhaled by the people living in the house. Depending on the exposure levels, time exposed, sensitivity to mold, age, and other factors, mold can aggravate and worsen asthma and lead to upper respiratory conditions such as:

- Nasal and sinus congestion
- Coughing, wheezing, and breathing problems
- Sore throat
- Skin and eye irritation
- Upper respiratory infections

If mold is suspected, you can inspect the basement for water damage or signs of excess moisture and other areas for water stains on walls, ceilings, and woodwork. The best tools to use for mold are your nose and eyes. If you see mold, or if you can smell an earthy, moldy smell, you can assume it is there. Testing for mold is considered by many experts to be a waste of money, money which could be better spent eliminating the problem.

Eliminating mold is a six-step process:

1. Identify the moisture source and eliminate it. Several possible sources have already been mentioned. Also check for good ventilation in basements and especially crawl spaces.

2. Dry all the wet areas and materials. You may need to move items around to allow better air flow. Use fans and dehumidifiers as necessary. Move wet items away from walls and off floors.

3. Remove contaminated materials and dispose of them properly. You should consider contracting Steps 3 and 4 to a firm which specializes in mold abatement. Specialized clothing, respirators, and other protective equipment should be worn when in contact with mold-laden materials. Care must be taken to prevent dislodging mold particles and spreading them to other areas. Hang plastic sheeting in doorways and bag up disposed materials.

4. Clean surfaces. Consider contracting this work, as it also requires the use of specialized equipment. Mold growing on non-porous surfaces can usually be cleaned, but safety considerations are of paramount importance. Check with your state Environmental Services Department for local requirements and recommendations.

5. Disinfect contaminated surfaces. If you wish, you can spray a mild bleach solution (1/2 cup bleach to 1 gallon of water) onto the areas where mold was removed to kill any mold missed by the cleaning. Wear skin protection and goggles to protect your eyes during the spraying. Get air moving across the area to prevent breathing the bleach fumes. Clean up any of the solution that runs off the area being cleaned but do not wipe the area you are disinfecting.

6. Keep a watch on the area to catch any signs of additional moisture which can lead to new mold growth. If moisture appears, repair the problem immediately.

For new construction, the concern is preventing mold in the future. The techniques for doing this include carefully installing vapor barriers under concrete slabs and in crawl spaces, making sure water drains away from the house, taking care when installing flashing, and careful checkout of the plumbing. Also be sure to insulate any lines and ducting which may sweat.

Termites

Most local building codes require treatment for termites as a part of the project if there is any new foundation work. This is normally done by having a licensed termite control company spray a heavy chemical barrier in the foundation trench. So long as this barrier is not disturbed, the company will guarantee protection from termites for a predetermined time period. For remodeling, if you expose old stud walls or are disturbing dirt around the foundation, get a termite inspection. These are relatively inexpensive and can prevent unpleasant future surprises.

Carbon Monoxide Testing

Carbon monoxide (CO) is a colorless, odorless, poisonous gas produced by the incomplete burning of solid, liquid, and gaseous fuels. Appliances fueled with natural gas, liquefied petroleum (LP gas), oil, kerosene, coal, or wood may produce CO. Burning charcoal and running cars produce CO. The symptoms of CO poisoning include headache, shortness of breath, dizziness, fatigue, and nausea. If you have appliances, heating, fireplaces, or portable heaters fired with any of the above, install CO detectors in your house. These should be in the room in which the appliances operate and in sleeping areas. CO detectors are about the same size and cost of a smoke detector and look a lot like a smoke detector. They will alarm before the CO reaches the danger level, giving you time to take action.

Chapter 15 Checklist

☐	If you are remodeling, check your home for radon. Mail-in test kits are available at most hardware superstores.
☐	If you are planning new construction and live in a radon high probability zone, check building codes for required or recommended preventive measures.
☐	Test for asbestos and either seal it or have it removed.
☐	Test older homes for lead-based paint, and seal or remove it.
☐	Check for mold in existing houses and take steps to eliminate moisture sources and existing mold colonies.
☐	Install vapor barriers and seal possible areas for leaks in new construction.
☐	Get a termite inspection if not current in existing houses, pre-treat ground for new construction.
☐	Install CO detectors if using fuels likely to generate CO.

CHAPTER 16

Potential Problems and Common Mistakes

The main reason you probably want to be your own contractor is to save money. The reality is that owner-builders, not being part of the building industry, sometimes pay more for materials than builders. You will have to buy or rent tools that the pros already have access to. Of course, it is possible to save money—and lots of it. Not every aspect of the project can be DIY. Knowing when to hire a contractor, and when you can step in and do the work yourself, will help you save money. There are numerous other small steps you can take to save on building costs.

The building process is complex, and unless the contractor (that is you!) has a great deal of knowledge about the construction business, mistakes can be costly. Many owner-builders lack technical knowledge and rely too much on suppliers or subs to give them information; some even purchase the wrong kind of material for the job, and end up paying twice for materials due to lack of leverage with the supplier.

Renovation projects that look easy on TV often turn into nightmares. With older homes, it is almost certain that each time you remove a wall or floor, you uncover something else that needs repair or replacement. There are also legal liabilities to consider.

Building or remodeling your own home can bring a deep sense of satisfaction and save you money. Everyone should try it at least once, as long as the answer to these two questions is yes:

- Do you have the time?
- Do you have the money?

Check your final design plan against this chapter before breaking ground. Avoiding the most common mistakes is at least a good start.

Designing a House That is Too Unique

The phrase "custom home" means that you are choosing to act as your own contractor because you want a house that is different from everyone else's. Building everything exactly as you would like it often turns out to be a big mistake. This is because if your taste differs too much from the norm, you will be stuck with a property that you cannot resell later. For most homeowners, the value of their home is a significant part of their investment portfolio.

I have seen a custom home with turquoise concrete and one with river rock and concrete walls; both these homes languished on the market. I watched an owner-builder enter an old, elegant house complete with a large ballroom and chop it up into so many rooms that within a year, he was financially drained, unable to sell, and filed for bankruptcy.

Your own mistake might not be as drastic as that, but building a house that is a great deal bigger or smaller than the others in the neighborhood can have nearly the same effect. Look at the existing homes around your property. If the neighborhood is not established yet, check with the homeowners' association to see what is being built. They will probably also have minimum and maximum square footage requirements — stay within those.

Financing

You can make several mistakes when financing a building project. One is applying for the construction loan too early. Although Chapter 4 advised learning all that you can about lenders far in advance, applying for a loan too early can be a waste of time. Rates change, loan packages change, and you may end up disclosing unnecessary information which can ultimately hurt your chances of getting a loan.

You can also waste money by applying too early. If you are not ready to submit the plans to your local building agency, the lender will not fund the loan. That is because he must wait on permits to be issued before he makes the funds available. In the meantime, you have put in an application, and your credit report and appraisal are going to expire. You are able to renew them, but it will cost some extra money. Apply for the loan about the same time you submit final plans to the building department — around 60 days before you plan to break ground.

Submitting Multiple Applications

When you make an application to a bank or a broker, they submit the loans to lenders. There are only about ten national lenders offering all-in-one loans (construction plus permanent). Your loan application might be submitted twice to the same lender through the two different bankers you have selected. If this lender sees any discrepancy in the information you have submitted, you could be turned down — and if he was the only lender who was perfect for your needs, you could end up without funding.

To prevent this, ask your loan officer who will fund the construction loan. Be sure that you are clear on the loan process, and that you stay on top of its progress. If there are problems, try to resolve them. And if the problems do not get resolved, do not feel obligated; just move on to the next loan officer.

Not Starting Out with Enough Funds to Finish

There is nothing worse than running out of money while building a home. It is impossible to know exactly what it will cost in the end. The costs of materials, labor, and the cost of delays are all prices that fluctuate.

Many people start building with their own cash savings, figuring that they will get a construction loan when the reserves have run out. This does not work for several reasons:

- If you are using credit cards to fund the project, your credit score may drop, making it difficult to get further funding.

- Bankers require a certain amount of cash reserves before they will make a loan. If you have put all of yours into the project already, they will not see this as viable cash reserves.

- The finishing work takes the bulk of the cost, so owners are lured into a false sense of security once the home is under roof. Then, when the cash runs low, they are unable to gather more funds, which makes the home unable to pass inspection.

To offset these problems, make sure you have financing firmly in place before you break ground. That will avoid all the risk associated with loan approval. Also, be sure that you borrow plenty. You can always pay extra money back, but it is nearly impossible to recover your losses on an underfunded project.

Bank Not Funding Draws

The bank sometimes does not fund draws when you have not completed the amount of building they feel is necessary to meet that percentage of the money. In other words, you have asked for 40 percent, but the bank only feels that you have completed 30 percent of the construction. This is difficult because you probably have subs waiting to be paid.

If this happens to you, first make sure that any sections that have been started are completed. Then meet with the bank, taking along your general contractor if you have one. Make sure that the bank understands what stage the project is in. Take along all paperwork, including lien releases (signed by subs). If you cannot convince the bank to supply the extra 10 percent, consider using your own cash or credit until the percentage of completion is clearly where it needs to be for the bank to release the funds.

Project is Taking Too Long

Problems will arise when building a house. The best planning and forethought cannot prevent some of the things that happen. How will you handle these problems?

Scheduling is really nothing more than an estimate. The weather takes over, materials are delayed, laborers do not show up, and your project is suddenly weeks behind. Rescheduling is difficult at best; a delay in one area has a domino effect, especially if subs are not available for other time slots. Choosing to use other subs rather than wait on the ones you had picked can be a risk because the workmanship may suffer.

If the construction loan is about to expire and the project is not complete, you have another problem. Most construction loans have a penalty if you go over the time limit. These can be costly – but they also can be negotiated.

As soon as you realize there is a scheduling problem, talk with the lender. Let him know exactly how much work is still needed to roll the construction loan into a permanent loan. Sometimes you can rearrange your work schedule, putting off jobs that are not necessary prior to the rollover. Then, if you need an extension, go ahead and negotiate just one (a single extension will cost you less than multiple extensions). If you are already past the deadline, the lender is more likely to charge the full penalty.

Subcontractors Did Not Finish the Job

This problem occurs more frequently than you might think. Your lender can tell you of projects that drag on for years because of people starting and not finishing.

This happens for a variety of reasons: subs can be unreliable, addicted to drugs or alcohol, or simply shady. Many are con artists who do this repeatedly as a way to make their living — that is why it is vital to check references. Sometimes you and the subs do not get along; even two good personalities can become incompatible when working together. Losing a sub can cost a great deal of time and money, and he could even file a lien against the property, tying you up for ages.

If the sub walks out, experts say the best thing to do is let it go. Some people work hard to get the sub back on track — one lady I know paid to send her carpenter's entire family to Disney World after he balked at refinishing (again) her hardwood floors in a different color. Even after the trip, he was sulky and the project took twice as long as it should have.

Another owner-builder paid his framers a full payment each week. At the end of two weeks, he had paid for two full weeks of work, even though the job amounted to a little over a week's worth. The framers then did not show up for three weeks — time he spent frantically trying to get in touch with them. This knocked the entire project off schedule. When he finally spoke to the lead framer, he heard, "Fire me, then." The sub had no reason to return to the job. The owner had to find a new sub.

If this happens to you, try to find a contractor who is willing to step in and finish the job. Take bids and examine his background just as you did the first contractor. Consider returning to the contractors you talked with before, to see if they have time in their schedule. You will probably have to pay a little more than you would like at this point, just to go forward and get the job done. You have to make compromises — like price and scheduling — just to get the house back on track.

Money Ran Out

Completely running out of money is dangerous when building a home, especially if you have already sold your existing dwelling. There are many reasons why people run out of funds. Be sure to read the final list in this chapter to help save costs. If you run out of money, here are a few quick fixes to try:

- **Use credit cards.** This is money you can pay back easily once construction is complete.
- **Borrow from a family member.** Ask for a short-term loan, and be willing to sign a promissory note. Do not sign over part ownership of the property — just promise to pay back the money on time.
- **Take out a home equity loan against another property.** Take out a home equity loan against the house you live in, or a vacation home. If you have not tapped into other properties, now is the time.
- **Borrow against other large assets.** Any asset you own, even stocks or jewelry, can produce a viable source for funding the project. Do not be afraid to put up assets for short-term solutions, but do not use this type of borrowing for long-term cost overruns.
- **Borrow against retirement accounts.** This does not mean liquidate them, although you could as a last resort. Instead, use a credit line that is secured against your account. Liquidating retirement accounts has other ramifications, like penalties and taxes, so check with an accountant before using this method to move the project forward.

This Project Is Putting Strain on My Marriage/Family

Any time-consuming project, like building a home, will add stress to your marriage and family life. The old adage "if you can last through building a house together, you can last through anything" is especially true if you are a DIY contractor.

Take this into consideration before you begin. Discuss the time factor with your spouse. You, your spouse, or both will have to devote 10 to 20 hours per week to the project until it is finished. Be clear about who will carry out which aspects of house building and home chores during the project.

During the project:

- Find ways to make it fun to work on together.
- Allow each other plenty of time to gripe and decompress every day.
- Understand the way your partner deals with stress and honor it.
- Take a time-out from building or renovation and focus on each other if necessary. Taking a vacation in a local hotel may seem like a luxury during a remodel.

The Lesser Evils

In addition to the major mistakes listed above, minor errors that can end up costing an owner-builder a great deal of money. Before deciding to build a home, be sure that you have addressed each of the following.

The lot

Before purchasing the lot, did you verify that it really is worth the price? Do not take a realtor or seller's word for the lot's value; get comparable prices and determine your own.

Restrictions

Many subdivisions are so full of restrictions that new owners find they cannot even live in the neighborhood, much less build a dream home there. Do not allow the HOA or review committee to tell you their covenants are "standard." There is no such thing as a standard set of covenants. Many neighborhoods have restrictions on building extra garages, landscaping, and even where you park your car at night. You also will not be allowed a broad interpretation of their covenant language, so be sure you let the committee interpret any restrictions that concern you.

Utilities

Utility hookups should be free, but if you extend beyond a certain distance (perhaps 150 ft.) the charge can be thousands of dollars. If you find a lot that is $25,000 less than the surrounding ones, ask yourself why.

Percolation (perc) tests

I have seen many buyers who do not understand the way perc tests work. Perc tests must be done according to the house they are planning to build and the exact location of the future septic system. A generic perc test has almost no value to you as a buyer. If the seller tells you a perc test has been done, ask to see its Health Department certificate. Make passing the test — your own test — a contingency of the sale.

Overgrown vegetation

When a lot is covered with dense growth, it is nearly impossible to evaluate the soil condition, view the rock formations, study the slope, or determine how it will drain. The cost of clearing a densely grown property can be substantial, especially if you are required to haul the waste to a landfill. Once cleared, the property might not be as suitable as once believed.

Spending too much time shopping

You can over-shop for many items. The first shopping frenzy is searching the Internet for the "perfect" set of plans, followed by a search for the "least expensive AND perfect" set of plans. Millions of plans available, and the Internet is a good place to get ideas. But the cost of an original set of plans created by a reputable designer might be very close to the price of one sold on the Internet. Many plans sold on the Internet do not provide all the details, so the subs are not able to use specific drawings for their work.

The next incessant shopping spree takes place in the bidding process. If you spend too much time digging for the lowest bid, you may also obtain the contractors who have trouble getting work due to poor work habits or sub-quality work. They might also cut their own costs by not buying insurance, not paying for licenses, and hiring illegal aliens — which can mean added costs for you.

Do not waste time looking for the lowest bid. Know the amount of work hours the job should take, and learn what a fair market price might be. Then get three to six good bids — and make a choice.

You can spend too much time shopping for materials. The ideal would be to find everything on sale someplace, but unless you are able to spend many days driving and searching you will not find all the lowest costs on materials. Instead of chasing down sales that are located six hours' drive away, work at building relationships with your own vendors or suppliers, who can extend you discounts and negotiate prices for you.

Getting It Wrong

As an inexperienced builder, you are bound to make some mistakes. You can avoid mistakes by knowing as much as you can before taking each step. For example, when you enter the local DIY store to buy supplies, you should already know exactly what materials you need. Do not rely on the (also inexperienced) store employees to recommend, say, mortar. They might suggest the wrong one for your particular job. Their mortar may require more drying time than you wanted, or allow cracks to form. Be sure that you know what you need, and that you get it.

Problem solving during the construction is an opportunity for subs to change orders or add to your contract. To avoid being taken advantage of, do not rely too much on the subs to create options. Each time a problem crops up, research it and find your own solution, then ask the sub for his input. That way you will know whether the solution he offers is viable.

Controls

Owner-builders notoriously underestimate time and costs. Often up to 10 percent of the job is missing from the master list. This renders the builder incapable of creating an efficient, cohesive plan. There is an optimal sequence of work for every job. This varies from one project to another. If things are done out of sequence, there may be additional work to pay for. Take the time to schedule very efficiently, and mentally run through your plan before putting it into action.

Conclusion

A Final Word about Saving Money

To save money, be alert at all times to cost-cutting solutions. There are some instances in which you do not want to scrimp.

- Buy the best lot you can afford.
- Hire the best contractors, especially for the foundation, framing, and carpentry.
- Do not sign a contract until you are sure that every detail is included, and you are satisfied with all the terms.

One of the best ways to save is to do work yourself. Offer to work as a helper on certain steps of the construction. Subs who do not have a regular helper will allow you to do this, and you will get on-the-job experience while you save on costs. Consider doing some of the other jobs, like:

- Painting
- Wallpaper
- Install doors and light fixtures
- Trim work
- Cleanup

When determining costs during pre-construction, always choose a bid over an estimate as a starting point. Using estimates can throw the budget way off, especially if you have forgotten something.

Consider downgrading materials that will not be seen; use prefabricated items if they are available, such as framing and windows with "fake" pane dividers.

Use a rectangular floor plan, carefully locating partitions to intersect the exterior wall studs to save on materials.

Cluster rooms that require plumbing fixtures, like bathrooms, kitchen, and utility rooms. For a two-story home, place the upper level plumbing directly over the lower level plumbing.

Consider using engineered floor trusses instead of standard lumber; eliminate double floor joists and band joists as well as bridging (which is no longer required for joists less than 2X12).

Select high-efficiency HVAC units with a high EER, and ENERGY STAR® appliances. These will save money in just a few months' bills.

In going over the final punch list, many small items might still be incomplete. There could be dings in the wall, paint that needs touching up, windows that need to be lined up, and so on. If your contract calls for the sub to come back and adjust or repair the work, be sure to get them back out soon. If not, these are items you can take care of yourself.

After completion, your property tax will go up almost right away. Homes are assessed according to value. Ordinarily, you have 60 days to challenge the assessment. Do not be unfriendly to the assessor, but be aware that you can refuse to allow them inside the home in many areas. When the assessment arrives, consider challenging the bill if comparable homes are appraising for less, or if your actual costs were less. Back up your claim with your receipts.

This book is meant to help you get started on the building process and give you confidence that you can build your own home. Good luck to you — and please feel free to write to me about your experience in building or remodeling your dream home!

APPENDIX A:

Resources

Websites

House plans and design

- Hanley Wood — **www.eplans.com**
- The Garlinghouse Company — **www.garlinghouse.com**
- Better Homes and Gardens — **www.houseplans.bhg.com**
- Building Green — **www.BuildingGreen.com**

Financing

- Indy Mac — **www.indymacmortgageservices.com**
- Construction Loan Center — **www.constructionloancenter.com**
- Custom Mortgage Corporation — **www.custommtgcorp.com**

Building

- **Do It Yourself — www.doityourself.com.** An Internet resource that gives practical advice on every sort of DIY project. It also offers links to suppliers and an online discussion forum.

- **For Residential Pros — www.forresidentialpros.com/building.** This online magazine is geared toward professional builders and architects, but it is full of good ideas for owner-builders, too.

- **Owner Builder Network — www.ownerbuildernetwork.com.** This website was created specifically for owner-builders, primarily Texas residents.

- **This Old House — www.thisoldhouse.com.** This Old House offers articles and a videos about every imaginable remodeling project.

Magazines

- *Architectural Digest*
- *Better Homes & Garden*
- *Do-it-Yourself*
- *Dwell*
- *HGTV*
- *Homebuilding & Renovating*
- *HOUSE*
- *Lowe's Creative Ideas*
- *SelfBuild and Design Magazine*
- *Taunton's Fine Homebuilding*
- *The Self Builder*
- *Timber Home Living*

Associations

The following associations are good resources for information on their members, as well as for locating professionals in the field:

- American Institute of Architects (AIA) — **www.aia.org**
- National Association of Home Builders (NAHB) — **www.nahb.org**

Other Good Stuff

These are materials that you probably will want to have on hand before beginning construction:

Your local building code — This is obtainable from the local building inspector. It outlines all the rules that must be adhered to when constructing or remodeling a home in your area.

HUD Guidelines to Minimum Property Standards (MPS) for Single Family Housing — This is especially helpful in areas of the country that have not adopted a building code. HUD requires that all properties insured with an FHA mortgage meet a code, so if you ever plan to sell the property it is good to have a copy of these codes. Obtain it from:

Department of Housing and Urban Development
FHA Standards, office of Manufactured Housing Programs
451 7th St. SW Room 9168
Washington, DC 20410-8000
Phone (202) 708-6423
http://portal.hud.gov/hudportal/HUD?src=/program_offices/housing/ramh/mps/mhsmpsp

APPENDIX B:

Sample Forms Checklist

The companion CD-Rom included with this book, contains the following sample forms, contracts, checklists, and timelines found throughout this book, plus more. The files are an easy-to-use format that you can edit or print off to assist while being your own contractor.

☐	Affidavit of Original or Sub-Contractor
☐	Before You Begin
☐	Bids Received Worksheet
☐	Building Inspection Approvals
☐	Categories for Your Notebook
☐	Change Order Form
☐	Checklist for Construction Loan Documents

☐	Checklist for Monitoring the Safety on Your Site
☐	Checklist for the Lot You are Considering
☐	Chapter Checklists
☐	Contractor Checklist
☐	Electrical Checklist
☐	Electrical Recheck
☐	Exterior Finish Checklist
☐	Exterior Finishing Checklist
☐	Job Site Safety Checklist
☐	Permit Checklist
☐	Plumbing Checklist
☐	Remodeling Checklist
☐	Decorating Needs
☐	Renovation Cost Breakdown Worksheet
☐	Sample Construction Timeline
☐	Sample Cost Breakdown
☐	Supplier Checklist
☐	Funding Options
☐	Key Items for Buildability Verification
☐	Kitchen Decisions to Make
☐	Lien Waiver Short Form
☐	Location Analysis
☐	Plumbing Fixtures

☐	Quality of Life Improvements
☐	Questions for Subcontractors
☐	Questions to Define the Factors
☐	Request for Proof of Insurance from Contractor
☐	Sample Independent Contractor Agreement
☐	Sample Independent Contractor Form
☐	Sample Subcontractor Bid Sheet
☐	Seven Steps to the Perfect Home Design
☐	Start-to-Finish Planning
☐	Supplier Contact Sheet
☐	Supplier Reference Sheet
☐	Take-Off List Information Sample Sheet
☐	The Six-Step Process to Eliminating Mold
☐	Things to Consider When Designing Your Home
☐	Three Levels of Features
☐	Tools That Will Keep You Organized

APPENDIX C:

House Building Glossary Terms

Abstract Of Title — A summary of recorded transactions concerning a property. It must be examined by an attorney or title company for any title defects which must be cleared before a buyer can obtain clear, marketable, and insurable title.

Acrylic Resin — A thermoplastic resin used in latex coatings

Acceleration Clause — Condition in a mortgage giving the lender the right to require immediate repayment of the loan balance if regular mortgage payments are not made, or for breach of other conditions of the mortgage

Accrued Interest — Interest which has been incurred but is not yet paid

Adjustable Rate Mortgage (ARM) — A mortgage in which the interest rate is adjusted periodically based on a pre-selected index. Subject to certain limitations, the rate and payments on an ARM loan rise and fall with the market.

Adjustment Interval or Adjustment Period — The length of time between rate adjustments on an Adjustable Rate Mortgage (ARM)

Aggregate – Irregularly shaped gravel suspended in cement

Agreement Of Sale — Contract signed by buyer and seller stating the terms and conditions under which a property will be sold

Air chamber – A pipe appendage with trapped air that is added to a line and serves as a shock absorber to slow or eliminate air hammer

Air-dried lumber – Dried via exposure to air without artificial heat. In the United States the minimum moisture content of thoroughly air-dried lumber is 12-15 percent; average is somewhat higher.

Alkyd resin – One of a large group of synthetic resins used in latex paint

Alternative Documentation — A substitute method of providing the documentation necessary to approve a loan, often used for people with hard-to-prove income. For example, they may use bank statements in lieu of pay stubs if it is not possible to provide written verification.

Amperage — The amount of current flowing in a wire

Amortization — The gradual reduction of a loan debt through periodic installment payments of principal and interest

Amortization schedule — A table showing the mortgage payment, broken down by interest and amortization, the loan balance, tax and insurance payments if made by the lender, and the balance of the tax/insurance escrow account

Anchor — Irons that hold together timbers or masonry

Anchor bolt — Bolt that secures a wooden sill plate to concrete or masonry floor or foundation wall.

Annual Percentage Rate (APR) — A calculation that expresses the total cost of a mortgage loan as a yearly rate (according to a federally mandated procedure). The APR calculation takes into account monthly interest payments, mortgage insurance, points, and certain fees paid at origination. It results in a rate slightly higher than the stated interest rate on the loan.

Apron — Trim that is used at the base of windows; also used as the base to build out crown molding

Appraisal — A written estimate of a property's current market value, based on recent sales information for similar properties, the condition of the property, and the neighborhood's impact on future property value

Appraisal Fee — A fee charged by a licensed, certified appraiser to provide an appraisal

APR — See Annual Percentage Rate

ARM — See Adjustable Rate Mortgage

Assessment — A local tax levied against a property for a specific purpose, such as road or sidewalk construction or sewer or street light installation

Asset Documentation — The documents that verify the existence of the borrower's assets

Assignment — The transfer of property rights by one person, the assignor, to another, the assignee

Assumability — A loan feature that allows the loan to be transferred from the seller to the purchaser of a new home with the same terms and conditions, subject to lender approval

Attic ventilators — in houses, screened openings that ventilate an attic space. They are located in the soffit area as inlet ventilators, and in the gable ends r along the ridge as outlet ventilators. They can also consist of power-driven fans used for an exhaust system.

BTU — British Thermal Unit. A standard unit of hot or cold air =output

Backfill — Process of placing soil against a foundation after all necessary foundation processes have been performed. Often, the foundation wall will need temporary interior bracing to support the weight of dirt, until it settles.

Backflow — The flow of water or other fluids into the main source of potable water

Backing — The bevel on the top edge of a hip rafter. It allows the roofing board to fit without leaving space between it and the lower side of the roof covering.

Balance Sheet — A document showing the financial situation — assets, liabilities, and net worth — of a company at a specific point in time

Balloon frame — The lightest and most economical form of construction. Studs and corner posts are continuous lengths from the first floor line or sill to the roof plate. The upper story floor joists are carried on ledgers or girts let into the studs.

Balloon mortgage — A short-term, fixed-rate loan with low payments for a set number of years and a large balloon payment of the remainder of the principal due at the end of the term

Baluster — A vertical support in a railing system; may be used on stairs, balconies, and porches

Balustrade — A railing system made of balusters, top rails, and possibly a bottom rail

Bankruptcy — Petition that may be brought by an individual or creditors, with a goal of orderly and equitable settlement of obligations. Proclamation must be given by a court of an individual's or organization's state of insolvency, or inability to pay debts.

Base shoe — Molding applied next to the floor on the interior baseboards, usually a quarter round strip

Batten — Narrow strips of wood covering joints, such as on plywood or wide board

Batter board — A pair of horizontally place boards nailed to vertical posts. Used at the corners of an excavation area to indicate desired level of excavation; also can be used for tying strings to indicate foundation wall perimeters.

Batter pile — Pile driven at an angle to brace a structure

Bazooka — An automated drywall application tool. It tapes, muds, and feathers in one pass

Bead — A corner or edge that must be finished with stucco

Bearer — The legal owner of a piece of property

Bed molding — The trim piece that covers the intersection of the vertical wall and any horizontal overhanging surface, like a soffit. In a series of moldings, the bed molding is the lowest one.

Bequest — A gift of personal property by will

Berm — A mound of built-up dirt used for either drainage or landscaping

Bevel board — Also called a pitch board, used in framing a roof or stairway to lay out bevels

Bibb — Special covering used where lines run through an exterior wall

Bill Of Sale — A document by which one transfers ownership of goods to another

Bi-weekly Mortgage — A payment plan under which the borrower pays one half of a monthly payment every two weeks

Blanket Mortgage — A mortgage covering at least two or more pieces of real estate, both of which together serve as collateral for the loan

Blind nailing — Nailing so that the nail heads are not visible on the face of the work. Used at the tongue of matched boards

Board foot — Unit of measure equal to a 1-in.-thick piece of wood one foot square. Length X Width X thickness = board feet

Boarding in — The process of nailing boards on the outside studs of a house.

Bolster — A short horizontal beam on top of a column. It supports and decreased the girder span. may be wood or steel.

Bond — A document representing a right to certain payments on underlying collateral

Boston ridge — A method of putting asphalt or wood shingles at the ridge or hip of a roof for a finished look

Break joints — A system of arranging joints so that they are not directly over adjoining joints. Done in shingles, siding, brickwork

Breather paper — Paper that lets water vapor pass through. Often used on outer walls to stop wind and rain while allowing water vapor to get out

Bridge Loans — Financing for homeowners who are constructing a new owner-occupied primary residence but are keeping the current owner-occupied residence, which they intend to sell

Broker — An independent middleman who arranges deals between borrowers and lenders. A broker is compensated for his services by the borrower and/or by the lender.

Budget — The budget is comprised of the total costs involved in the construction of the new home, the amount of costs that the borrower may have already paid, and the costs remaining to be paid to complete the home. This will quickly identify the borrower's equity already paid, and the amount of borrower funding needed at closing, if any.

Building paper — Inexpensive thick paper used to insulate a building prior to putting roofing on. Sometimes placed between each level of the home

Built-up roof — A roof that is composed of 3 to 5 layers of asphalt felt, laminated with coal tar or asphalt. The top is crushed slag or gravel

Buy-Down — A situation in which the seller of a property contributes money, allowing the lender to give the buyer a lower rate and payment, typically in exchange for an increase in sales price

Buyer's broker — An agent hired by a buyer to locate a property for purchase and to represent the buyer in negotiations with the seller's broker

Buyers' Market — Market conditions that favor buyers. With more sellers than buyers in the market, buyers have ample choice of properties and can negotiate lower prices.

Call option — A loan feature in which the lender may require repayment of the loan in full before the term of the loan is up

Cant strip — A piece of lumber used at the junction of a deck, or roof, and a wall to modify the angle. It eases or eliminates the effect of a sharp angle to reduce the possibility of cracking.

Cantilever — A horizontal structure component that projects from a building, such as a step, balcony, beam or canopy, that is without external bracing and appears to be self-supporting

Cap (construction) — Hardware that terminates plumbing lines; or the upper member of a column, door cornice, or molding

Caps (Interest) — Consumer safeguards which limit the amount the interest rate on an adjustable rate mortgage can change in an adjustment interval and/or over the life of the loan

Caps (Payment) — Consumer safeguards which limit the amount monthly payments on an adjustable rate mortgage may change at a given time

Casement window — A window that swings out to the side on hinges

Cash out — A refinance for more than the balance of the current mortgage to take cash out from the loan at closing

Casing nails — Used to apply finish trim and millwork; the nail head is set below the surface of the wood so that it is hidden.

CC&Rs — See Covenants, Conditions and Restrictions.

Ceiling — The maximum allowable interest rate of an adjustable rate mortgage

Certificate Of Eligibility — Document issued by the Veterans Administration to qualified veterans which entitles them to VA-guaranteed loans

Certificate Of Occupancy — Document issued by local government agency stating that a property meets the requirements of health and building codes

Certificate Of Reasonable Value (CRV) — A property appraisal performed by a VA-approved appraiser establishing the limit on the principal of the VA loan

Certificate Of Title — Written opinion of the status of title to a property, given by an attorney or title company. This certificate does not offer the protection given by title insurance

Certificate Of Veteran Status — Document given to veterans or reservists who have served 90 days of continuous active duty (including training time) which enables them to obtain lower down payments on certain FHA/VA-insured loans

Chain of Title — The chronological order of conveyance of a property from the original owner to the present owner

Chamfer — The beveled edge of a board

Checks — Splits or cracks in a board; usually caused by seasoning

Clean-out — A sealed opening in a pipe that can be removed to clean out a clog.

Clear Title — Real property ownership free of liens, defects, encumbrances or claims

Cleat — A length of wood affixed to a surface to give a firm foothold, or to hold an object in place

Closing (or Settlement) — Meeting between the buyer, seller, and lender or their agents, at which property ownership and funds legally change hands

Closing Agent — Neutral third party appointed to act as a custodian for documents and funds during the transfer of property from seller to buyer. Depending on local law and custom, this could be an attorney, escrow agent, or title company.

Closing Costs — Costs associated with the closing of the loan (title costs, loan fees, discount fees, inspection fees, appraisals)

Closing/Settlement Statement — A form prepared by the closing agent that itemizes the closing costs associated with purchasing or refinancing a home. Also called HUD-1

Cloud on Title — An outstanding claim or encumbrance that, if valid, would affect or impair the owner's title

Combined Loan To Value (CLTV) — The percentage of the property value borrowed through a combination of more than one loan. The combined loans and lines of credit divided by property value equals Combined Loan To Value Ratio.

COFI — See Cost of Funds Index

Collar beam — Boards 1 or 2 in. thick connecting roof rafters near the ridge board in order to strengthen the roof structure

Collateral — Assets that secure a loan (In the case of a mortgage, real property serves as collateral.)

Conforming Loan — A mortgage loan eligible for purchase by the two federally sponsored housing agencies, Fannie Mae and Freddie Mac

Construction APR — A calculation that expresses the cost of a mortgage loan as a yearly rate (according to a federally mandated procedure) over the life of the loan, including the construction phase.

The APR calculation takes into account monthly interest payments, mortgage insurance, points, and certain fees paid at origination. It often results in a rate higher than the stated interest rate on the Note, as well as the estimated APR disclosed on the permanent financing phase of the loan term. You may receive two APRs, one for the construction period of your loan and the other for the permanent financing of your loan, or the APR can be combined for both the construction and permanent periods of your loan.

Construction costs — Costs to complete the construction of a new home: off-site, on-site, land value, closing costs, contingency, and interest reserves

Construction loan — A short-term interim loan to fund the construction of buildings or homes, which usually advances the money in installments as work progresses

Construction-to-permanent loan — A bank loan that allows the owner-builder to make draws during the construction phase, then convert to a fixed-rate loan once construction is completed without having to refinance the property

Contingency — A condition which must be satisfied before a contract is legally binding — before a sale can close

Conventional Loan — A mortgage not insured by the FHA or guaranteed by the VA

Conversion Clause — A provision in some ARMs that allows an ARM to convert to a fixed-rate loan, normally after the first adjustment period. The new fixed rate is based on a formula tied to current rates. There may be a charge for the conversion feature.

Convertible ARMs — ARMs with the option of conversion to a fixed loan during a given time period

Conveyance — The transfer of a deed, lease, or mortgage

Corbel — Extending a course of bricks beyond the face of a wall. The total projection should never exceed the thickness of the wall.

Cornice — Trimwork finishing the intersection of the roof with the siding; consisting of a fascia board, a soffit, and appropriate moldings

Cornice return — The underside of the cornice; trim, transitioning from the horizontal eave line to the sloped roofline

Cost Of Funds Index (COFI) — A common index used in adjustable rate loans based on the weighted-average interest rate paid for deposits by savings institutions that are members of the 11th Federal Home Loan Bank District

Course of construction — Insurance policy in the form of an "all risk" policy with fire, extended coverage, builder's risk, replacement cost, vandalism, and malicious mischief insurance coverage. The owner is named insured with insurable value equal to the replacement cost of the improvement or the loan amount, whichever is lower. Once the improvements are completed and the permanent mortgage begins, the course of construction policy is usually converted to a standard "all risk" policy.

Covenants, Conditions, and Restrictions (CC&Rs) — A document that defines the use, requirements and restrictions of a residential area, condominium or Planned Unit Development (PUD)

Countersink — to set the head of a nail below the surface

Cove molding — Has a concave face; used as trim or for finishing interior corners

Crimp — A crease formed in sheet metal to make it less flexible or to make it easy to fasten onto

Cripple — A stud used for bracing under structural framing, like windows

Crown molding — trim piece applied at the intersection of ceiling and walls in formal rooms

Crusher run — crushed stone with sharp edges, used for driveway and foundation support as a base. Packs down to give a stable surface

Cube — A standard ordering unit for masonry block: 6X6X8

Dado — A rectangular groove across the width of a board or plank; ornamental

Debt-To-Income Ratio — A ratio that compares the total amount of recurring debt payments a borrower is obligated to make to the amount of their income

Deed — Legal document by which title to a property is transferred from one owner to another. The deed contains a description of the property and is signed, witnessed, and delivered to the buyer at closing

Deed Of Trust — Document creating a lien on a property as security for the payment of a debt. In some states, a mortgage is used instead

Default — Failure to meet legal obligations in a contract, including failure to make payments on a loan. A mortgage is considered to be in default when a payment is 30 days past due.

Deferred Interest — Amount added to the balance of a loan when monthly payments are insufficient to cover the interest incurred. This results in negative amortization.

Delinquency — Failure to make required payments on time

Deposit — Cash paid to the seller when a formal sales contract is signed

Depreciation — Decline in property value

Documentary stamps — A state tax, in the form of stamps, required on deeds and mortgages when real estate title passes from one owner to another

Door jamb — The surrounding case into which a door closes. It consists of two upright side jambs and a horizontal head jamb

Dormer — An opening in a sloping roof that projects out to form a wall suitable for windows or other openings

Dovetail — Joint made by cutting pins in the shape of dove tails which fit between matching cuts on another piece of wood

Draw — During construction, a request for a percentage of the funds by the builder. The bank holds the funds and disburses the money in draws.

Earnest money — Deposit made by a buyer toward the down payment as evidence of good faith when the purchase agreement is signed

EER — See Energy Efficiency Rating.

Eaves — the part of the roof that extends beyond he outside walls of a house.

Effective Interest Rate — The cost of a mortgage expressed as a yearly rate, usually higher than the interest rate on the mortgage since this figure factors into the up-front costs of acquiring the loan

Encumbrance — Any lien, such as a mortgage, tax or judgment lien. It can also be an easement a restriction on the use of the land or an outstanding dower right that may diminish the value of the property.

Energy Efficiency Rating — A national rating system that must be displayed on appliances that measures the efficient use of electrical power

Equity — The difference between the current market value of a property and the outstanding mortgage balance

Equity loan — A loan based on the borrower's equity in his or her home

Escrow — An account set up by the lender into which the borrower makes periodic payments for taxes, hazard insurance, assessments, and mortgage insurance premiums

Estimated Settlement (or Closing) Statement — A document provided by the closing agent a few days before closing, detailing all costs and indicating the final sum the buyer will be required to bring to the closing

Expansion joint — A bituminous fiber strip that separates blocks or concrete to allow for expansion (due to temperature changes) without cracking

Face nailing (direct nailing) — Nailing perpendicular to the board or other material

Fascia — A flat board covering the end of the rafter, or the board that connects the top of siding to the bottom of the soffit. Also the board of the cornice to which the gutter is attached

Fannie Mae (FNMA) — Corporation created by Congress that buys and sells residential mortgages. Fannie Mae provides funds for one in seven mortgages.

Farmer's Home Administration (FmHA) — An agency of the US Department of Agriculture that provides financing for purchasers of new homes and farms in small towns and rural areas

Federal Deposit Insurance Corporation (FDIC) — Independent deposit insurance agency created by Congress to maintain stability and public confidence in the nation's banking system

Feathering — Applying drywall compound in successive coats, thereby widening the compound joint

Federal Housing Administration (FHA) — Government agency, division of the Department of Housing and Urban Development, which insures residential mortgage loans made by private lenders and sets standards for underwriting mortgage loans

Federal Reserve — Central bank of the United States and major regulatory agency for many commercial banks

Fee Simple — Absolute ownership of real property

Ferrule — Aluminum sleeve used to attach trough to gutter spike

Fill dirt — Loose dirt brought in from another location to use under slabs, driveways, and sidewalks; sturdier than topsoil

Finish grade — The process of leveling and smoothing topsoil into its final position

Fire stop — A solid, tight closure of a concealed space in order to prevent fire from spreading. In a frame wall, this usually requires 2X4 cross blocking between studs

Fishplate — A wood or plywood piece used at the junction of opposite rafters near the ridge line.

First mortgage — The primary lien against a property

Fixed-rate mortgage — A mortgage whose interest rate does not change for the life of the loan. Payments are also fixed

Flood insurance — A form of hazard insurance required by the federal government to cover property damage or loss in flood zones

Floor — The minimum interest rate payable on an Adjustable Rate Mortgage

FICO Score — A credit evaluation score developed by Fair, Isaac, and Co., used by lenders as one factor in making a loan decision. Some methods of improving a score are to establish and maintain a payment history on credit accounts, keep public records (bankruptcies, judgments, etc.) and collection accounts to a minimum, pay down loans, keep credit cards well below their limits, avoid late payments, and limit applying for new credit.

Fixed Price Contract — A construction contract between the borrower and contractor defining the cost of building and improving a residence. The contract should have a start date and a finish date.

Flashing — Galvanized sheet metal used for a lining around joints, normally between shingles and chimneys, exhaust or ventilation vents, and other protrusions that might allow water seepage under the shingles

Flue — The passage in the chimney through which smoke, gas, and fumes ascend

Forbearance — Grace period given when a lender postpones foreclosure to give the borrower time to catch up on overdue payments

Freddie Mac (FHLMC) — Agency that purchases conventional mortgages from insured depository institutions and HUD-approved mortgage bankers

Furring — Long strips of wood attached to walls or ceilings to allow attachment of drywall or ceiling tiles. "Furring out" refers to using the furring strips to bring the wall further into the room. Furring down refers to using them to lower a ceiling.

GFCI — See Ground-Fault Circuit Interrupter.

Gambrel — A roof that slopes steeply at the edge of a building, changing to a shallower slope at the center

Gem box — metal box installed during the electrical rough-in that holds outlets and receptacles

General Liability — A comprehensive general insurance policy or a broad form liability endorsement provided by the borrower or the general contractor

Government National Mortgage Association (GNMA, or Ginnie Mae) — A federal association working with the FHA that offers special assistance in obtaining mortgages and purchases mortgages in the secondary market

Good faith estimate — Written estimate of costs the borrower will pay at closing, provided by a lender within three days of loan application

Graduated Payment Mortgage (GPM) — Mortgage in which initial low payments (with potential negative amortization) increase regularly for several years and then level off

Grace Period — Period of time during which a loan payment may be made after its due date without incurring a late penalty

Gross Income — Total income before taxes or expenses are deducted

Ground-Fault Circuit Interrupter — An extra-sensitive circuit breaker used for additional protection against shock

Guarantee or Guaranty — A promise by one party to pay a debt or perform an obligation contracted by another in the event of that person's default.

Gusset — a flat wood, plywood, or similar piece providing a connection at the intersections of wood. Usually used in the joints of wood trusses.

Gypsum board — Drywall

Gypsum plaster — A mix of gypsum that is made to be sued with sand and water added for base coating plaster

H clip — A metal clip for holding adjacent plywood sheets in alignment

Hazard Insurance — A policy that protects the insured against loss due to fire or certain natural disasters in exchange for a premium paid to the insurer. Also known as Home Owner's Insurance or fire insurance

Head lap — Vertical length in inches of the amount of overlap between shingles

Header — One or two pieces of lumber placed over doors and windows to support the load above them

Heel of a rafter — End or foot that rests on the wall plate

Heel wedges — Triangular shaped wood that is driven into gaps between rough framing, like window framing, to give solid backing

Hip roof — A roof sloping upward toward the center from every side; requires a hip rafter at each corner.

Home Equity Loan — An additional mortgage secured by the equity in the home. All funds for this loan are disbursed at closing.

Home Equity Line Of Credit — A revolving line of credit secured by the equity in the home. Unlike a Home Equity Loan, these funds may be drawn and repaid like a credit card.

Homeowner's Warranty — A type of insurance that covers repairs to specified parts of a house for a specific period of time

Hopper window — window that is hinged at the bottom and swings inward

Housing and Urban Development (HUD) — A U.S. government agency established to implement federal housing and community development programs; oversees the Federal Housing Administration.

Housing Code — Local government ordinance that sets minimum standards of safety and sanitation for existing residential buildings

HUD-1 Settlement Statement — A form mandated by the federal government that itemizes the closing costs associated with purchasing a new home. Also called Estimated Settlement Statement

I-beam — steel beam with a cross section so that it is shaped like a capital I; used for long spans over wide openings where there are heavy roof loads imposed above the opening

Index — A published rate used by lenders to calculate interest adjustments on adjustable rate mortgages (Index + Margin = Interest Rate). Common indexes include 1-Year Treasury securities, COFI (Cost Of Funds Index), and Six-Month LIBOR (London Interbank Offered Rate)

Independent Contractor — A person who, in performing services for another, is responsible only for the final result, and is not subject to control as to the methods used to achieve that result

Initial Rate — The rate charged during the first interval of an adjustable rate mortgage

Insulated Block — Hollow block masonry filled with insulation

Insulation — Any material used to reduce the effects of heat, cold, or sound transmission and to reduce fire hazard.

Any material used in the prevention of the transfer of electricity, heat, cold, moisture, and sound

Insulative panel — Flat sheet material made of insulative material to improve thermal resistance characteristics of assembly

Insolvency — Condition of a person unable to pay debts as they fall due

Interest — Charge paid for borrowing money

Interest Rate — The rate, expressed as a percentage, of the outstanding balance used to calculate interest charges

Interest Rate Cap — A safeguard built into ARMs to prevent drastic changes in interest rates

Interest Reserve — During the construction period, an account is established to pay the estimated interest costs during the construction of the new home. Since the borrower is only charged interest on the amount of funds disbursed, an estimate of the average disbursed amount is made. The bank estimates that, on average, a percentage of the loan amount will be disbursed during the term of the construction period.

Investor Rehab Financing — This program is designed to provide a loan for investors to acquire and rehabilitate a property for future rental use.

Involute — Curved part of trim that terminates a piece of staircase railing in traditional homes

Isolation joint — A joint in which two incompatible materials are isolated from each other to prevent chemical action between the two

Jack rafter — A shortened rafter that joins a hip or valley to the top of a wall plate

Jack post — A hollow metal post with a jack screw in one end to adjust the height.

Jack stud — A short stud that does not extend from floor to ceiling; may reach from floor to a window, for example

Jalousie window — A window with stationary or adjustable blinds angled to permit air and provide shade, while at the same time preventing rain from entering

Jamb — The vertical side of a door or window. The side piece or post of an opening; commonly applied to the door frame

Joint liability — Liability shared among two or more people, each of whom is liable for the full debt

Joint tenancy — The ownership of property by two or more persons with the survivor taking the share of the deceased

Joint — The point of connection between structural members

Joint clip — A fastener used vertically, sharp edges down, over the edges of two pieces, and then hammered down into them. In plywood sheathing, the clip fastens two abutting pieces of plywood

Joist — One of a group of light, closely spaced beams used to support a floor deck or flat roof. Timbers supporting the floorboards

Joist hanger — A metal stirrup that supports the ends of joists so that they are flush with the girder

Joule — In the international system of units, the amount of energy needed to raise one kilogram of water one degree Celsius

Journeyman — An experienced reliable worker who has learned his trade and works from another person.

Junction Box — A metal box in which runs of cable meet and are protectively enclosed

Kerf — The area of a board removed by the saw; vertical notch or cut in a batter board where a string is fastened

K Bracing — A form of bracing where a pair of braces located on one side of a column terminates at a single point within the clear column height

Key — Fancy decorative lintel above window made of brick; also called keystone.

Keyways — Tongue and groove connection where perpendicular concrete components meet; designed to prevent movement between components

Keyed joint — A joint in which one structural member is keyed or notched into an adjoining member as in timber construction. In masonry construction, a finished joint of mortar which has been tooled concave

Kick plate — A metal strip or plate that runs along the bottom edge of a door to protect against the marring of the finished surface

Kiln dried — Lumber dried by means of controlled heat and humidity

Kip — A unit of weight or force equal to 1,000 lbs

Knee brace — A corner brace, fastened at an angle from wall stud to rafter, stiffening a wood or steel frame to prevent angular movement

Lag screws — Large screws with heads designed to be turned with a wrench

Laminate — Thin material, often plastic or wood, glued to the exterior of a cabinet

Landing — A platform between flights of stairs

Lath — A grid applied to exterior sheathing used as a base for stucco

Lender's Contingency — This is a reserve to cover unforeseen circumstances in the construction of the new home. At a minimum, 5 percent of the "on-site costs" will be established in the contingency account (separate from the contractor's).

Lien — A legal claim against a property that must be paid when the property is sold

Lifetime Interest Rate Cap — The highest interest rate that can be charged for an adjustable rate mortgage during the life of the loan

Lintel — The horizontal beam placed over an opening

Loan Origination Fee (or Processing Fee) — Fee charged by a lender that compensates for the work in evaluating and processing the loan

Lock (or Lock In) — A lender's guarantee of an interest rate and related points for a set period of time, usually between loan application and loan closing. Protects borrower against rate increases during that time

Loan To Value (LTV) Ratio — The percentage of the property value borrowed (loan amount/property value = loan to value ratio)

LTV — See Loan To Value Ratio.

Luminaire — A complete lighting unit consisting of a light source, switch, globe, reflector, housing, and wiring

Mansard roof — A type of roof that slopes very steeply around the perimeter of the building to full wall height. The center of it is either flat or very low sloped.

Margin — The percentage amount added to an index to calculate the interest rate of an adjustable rate mortgage at each adjustment

Marketable title — A title that is free and clear of liens, clouds, or other defects which would prevent the sale of the property.

Market value — The value that a willing seller would accept and a willing buyer would offer given a reasonable time for the seller to market a property.

Mastic — Pasty material used for setting tile or for protective coating for thermal insulation or waterproofing. Comes in caulk tubes or 5-gallon cans.

Millwork — Generally, all materials made of finished wood (doors, windows, door frames, blinds, moldings) and manufactured in millwork plants are called millwork

MIP (Mortgage Insurance Premium) — Insurance purchased by borrower to insure against default on government (FHA or VA) loans

Moisture barrier — A membrane used to prevent the migration of liquid water through a floor or wall

Moisture movement — The movement of moisture through a porous medium; the effects of such movement on efflorescence and volume change in hardened cement paste, mortar, concrete or rock

Mono Pitch Truss — A truss that would develop a shed type roof

Mortgagee — The lender in a mortgage loan transaction

Mortgage — Document creating a lien on a property as security for the payment of a debt. In some states, a Deed of Trust is used instead.

Mortgage banker — A lender that originates and funds, then sells and services mortgage loans

Mortgage broker — A person or entity that arranges financing for borrowers, but places loans with lenders rather than funding them with the broker's own money

Mortgage insurance — Insurance purchased by a buyer to cover the lender's risk of loss. Mortgage Insurance is generally required by lenders when the down payment is less than 20 percent of the purchase price.

Mortgage loan — A loan for which real estate serves as collateral to provide for repayment in case of default

Mortgage note — Legal document obligating a borrower to repay a loan at a stated interest rate during a specified period of time. The agreement is secured by a mortgage.

Mortgagor — The borrower in a mortgage loan transaction

Mortar — A mixture of cement paste and fine aggregate; in fresh concrete, the material occupying the interstices among particles of coarse aggregate; in masonry construction, mortar may contain masonry cement, or may contain hydraulic cement with lime to afford greater plasticity and workability than are attainable with standard hydraulic cement mortar. A substance used to join masonry units consisting of cementitious materials, fine aggregate, and water

Mortise — A slot cut edgewise into a board to receive a tenon of another board to form a joint

Mosaic — Tile with small inlaid pieces of porcelain or natural clay materials to form decorative patterns. Small tile or bits of tile, stone, or glass used to form a surface design or an intricate pattern

Mullion — a vertical divider in the frame between windows, doors, or other openings

Muntin — a short bar separating glass panes in a window sash

Nail inspection — An inspection made by a municipal building inspector after the drywall material is hung with nails and screws (and before taping)

Natural finish — A transparent finish which does not seriously alter the original color or grain of the natural wood. Natural finishes are frequently provided by sealers, oils, varnishes, water repellent preservatives, and other similar materials.

NEC (National Electrical Code) — A set of rules governing safe wiring methods. Local codes—which are backed by law—may differ from the NEC in some ways.

Negative Amortization — A condition created when a loan payment is less than interest alone. Even though payments are made on time, the amount owed increases.

Neutral wire — Color-coded white, this carries electricity from an outlet back to the service panel. Also see hot wire and ground

Newel post — The large starting post to which the end of a stair guard railing or balustrade is fastened

Nonbearing wall — A wall supporting no load other than its own weight

Non-Assumption Clause — A statement in a mortgage contract forbidding the assumption of the mortgage by another borrower without the prior approval of the lender

Non-dischargeable Debt — Debt, such as taxes, that cannot be forgiven in a bankruptcy liquidation

Notice Of Default — Written notice to a borrower that a default has occurred

and that legal action may be taken

Nosing — The projecting edge of a molding or drip or the front edge of a stair tread

Notch — A crosswise groove at the end of a board

Note — A formal document showing the existence of a debt and stating the terms of repayment

Nozzle — The part of a heating system that sprays the fuel of fuel-air mixture into the combustion chamber

Note — Legal document stating the terms of a debt and a promise to repay it

O.C. (On Center) — The measurement of spacing for studs, rafters, and joists in a building from the center of one member to the center of the next.

Oakum — Loose hemp or jute fiber that is impregnated with tar or pitch and used to caulk large seams or for packing plumbing pipe joints

Off-Site costs — These are indirect site costs. Permit fees, engineering fees, architectural fees, and other costs associated with building the new home but not directly a part of the actual construction costs

Open hole inspection — When an engineer (or municipal inspector) inspects the open excavation and examines the earth to determine the type of foundation (caisson, footer, wall on ground) that should be installed in the hole.

Oriented Strand Board (OSB) — A manufactured 4 ft. X 8 ft. wood panel made out of 1 in.- 2 in. wood chips and glue. Often used as a substitute for plywood

Outrigger — An extension of a rafter beyond the wall line. This is often a smaller member nailed to a larger rafter to form a cornice or roof overhang

Outside corner — The point at which two walls form an external angle, one you as a rule can walk around

Overhang — Outward projecting eave-soffit area of a roof; the part of the roof that hangs out or over the outside wall. See also Cornice.

Pad out, pack out — To shim out or add strips of wood to a wall or ceiling in order that the finished ceiling/wall will appear correct

Pallets — Wooden platforms used for storing and shipping material. Forklifts and hand trucks are used to move these wooden platforms around.

Panel — A thin flat piece of wood, plywood, or similar material, framed by stiles and rails as in a door (or cabinet door), or fitted into grooves of thicker material with molded edges for decorative wall treatment

Paper, building — A general term for papers, felts, and similar sheet materials used in buildings without reference to their properties or uses. Normally comes in long rolls

Paper, sheathing — paper or felt used in wall and roof construction to protect against air and moisture passage

Parapet —A wall placed at the edge of a roof to prevent people from falling off

Parting stop or strip — A small wood piece used in the side and head jambs of double hung windows to separate the upper sash from the lower sash

Particle board — Plywood substitute made of course sawdust that is mixed with resin and pressed into sheets. Used for closet shelving, floor underlayment, stair treads, etc.

Partition — A wall that subdivides spaces within any story of a building or room

Paver, paving — Materials—commonly masonry—laid down to make a firm, even surface

Payment cap — Limit on the amount by which a borrower's adjustable rate mortgage payments may increase, regardless of rise in interest rates. May result in negative amortization

Payment schedule — A pre-agreed upon schedule of payments to a contractor usually based upon the amount of work completed. Such a schedule may include a deposit prior to the start of work. There may also be a temporary 'retainer' (5-10 percent of the total cost of the job) at the end of the contract for correcting any small items which have not been completed or repaired.

Pedestal — A metal box installed at various locations along utility easements that contain electrical, telephone, or cable television switches and connections

Penalty clause — A provision in a contract that provides for a reduction in the amount otherwise payable under a contract to a contractor as a penalty for failure to meet deadlines or for failure of the project to meet contract specifications

Penny — As applied to nails, it originally indicated the price per hundred. The term now series as a measure of nail length and is abbreviated by the letter "d". Normally, 16d (16 "penny") nails are used for framing

Per Diem Interest — Interest calculated per day. Depending on the day of the month on which closing takes place, borrower pays interest from the date of closing to the end of the month. The first mortgage payment of a loan is generally due on the first of the following month.

Periodic Interest Rate Cap — A limit on the amount that interest rates can change at each adjustment period

Percolation test or perc. test — Tests that a soil engineer performs on earth to determine the feasibility of installing a leech field type sewer system on a lot. A test to determine if the soil on a proposed building lot is capable of absorbing the liquid affluent from a septic system

Performance bond — An amount of money (10 percent of the total price of a job) that a contractor must put on deposit with a governmental agency as an insurance policy that guarantees the contractors' proper and timely completion of a project or job

Perimeter drain — 3-in. or 4-in. perforated plastic pipe that goes around the perimeter (either inside or outside) of a foundation wall (before backfill) and collects and diverts ground water away from the foundation. It is "daylighted" into a sump pit inside the home, and a sump pump is sometimes inserted into the pit to discharge any accumulation of water.

Permeability — A measure of the ease with which water penetrates a material

Permit — A governmental municipal authorization to perform a building process such as a zoning\use permit, demolition permit, grading permit, septic permit

Pigtails, electrical — The electric cord that the electrician provides and installs on an appliance such as a garbage disposal, dishwasher, or range hood

Pier — A column of masonry, usually rectangular in horizontal cross section, used to support other structural members

Pigment — A powdered solid used in paint or enamel to give it a color

Pilot hole — A small-diameter, pre-drilled hole that guides a nail or screw

Pilot light — A small, continuous flame (in a hot water heater, boiler, or furnace) that ignites gas or oil burners when needed

Pitch — The incline slope of a roof or the ratio of the total rise to the total width of a house. Roof slope is expressed in the inches of rise, per foot of horizontal run

PITI — Principal, interest, taxes and insurance (the four major components of monthly housing payments)

Plan view — Drawing of a structure with the view from overhead, looking down.

Plate — Normally a 2 X 4 or 2 X 6 that lays horizontally within a framed structure, such as:

Sill plate — A horizontal member anchored to a concrete or masonry wall

Sole plate — Bottom horizontal member of a frame wall

Top plate —Top horizontal member of a frame wall supporting ceiling joists, rafters, or other members

Plenum — The main hot-air supply duct leading from a furnace

Plot plan — An overhead view plan that shows the location of the home on the lot. Includes all easements, property lines, setbacks, and legal descriptions of the home. Provided by the surveyor

Plough, plow — To cut a lengthwise groove in a board or plank. An exterior handrail normally has a ploughed groove for hand gripping purposes

Plumb — Exactly vertical and perpendicular

Plumb bob — A lead weight attached to a string. It is the tool used in determining plumb

Plumbing boots — Metal saddles used to strengthen a bearing wall/vertical stud(s) where a plumbing drain line has been cut through and installed

Plumbing ground — The plumbing drain and waste lines that are installed beneath a basement floor

Plumbing jacks — Sleeves that fit around drain and waste vent pipes at, and are nailed to, the roof sheeting

Plumbing rough — Work performed by the plumbing contractor after the Rough Heat is installed. This work includes installing all plastic ABS drain and waste lines, copper water lines, bath tubs, shower pans, and gas piping to furnaces and fireplaces. Lead solder should not be used on copper piping.

Plumbing stack — A plumbing vent pipe that penetrates the roof

Plumbing trim — Work performed by the plumbing contractor to get the home ready for a final plumbing inspection. Includes installing all toilets (water closets), hot water heaters, sinks, connecting all gas pipe to appliances, disposal, dishwasher, and all plumbing items

Plumbing waste line — Plastic pipe used to collect and drain sewage waste

Ply — A term to denote the number of layers of roofing felt, veneer in plywood, or layers in built-up materials, in any finished piece of such material

Plywood — A panel (normally 4 ft. X 8 ft.) of wood made of three or more layers of veneer, compressed and joined with glue, and most often laid with the grain of adjoining plies at right angles to give the sheet strength

Point load — A point where a bearing/structural weight is concentrated and transferred to the foundation

Points (or Discount Points) — Money paid to a lender at closing in exchange for a lower interest rate. Each point is equal to 1 percent of the loan amount.

Portland cement — Cement made by heating clay and crushed limestone into a brick and then grinding to a pulverized powder state

Post — A vertical framing member typically designed to carry a beam. Often a 4X 4, a 6X 6, or a metal pipe with a flat plate on top and bottom

Post-and-beam — A basic building method that uses just a few hefty posts and beams to support an entire structure

Power Of Attorney — Legal document authorizing one person to act on behalf of another

Prepaid Expenses — Taxes, insurance, and assessments paid in advance of due dates

Prepaid Interest — Interest charged to a borrower at closing to cover interest on the loan between closing and the end of the month in which the loan closes

Prepayment — Full or partial payment of the principal before the due date. This might occur if the borrower makes extra payments, sells the property, or refinances the existing loan

Prepayment Penalty — Fee that may be charged by a lender for early payment of debt

Prequalification — The process of estimating how much money a prospective new homebuyer will be eligible to borrow prior to application for a loan

Premium — Amount payable on a loan

Preservative — Any pesticide substance that, for a reasonable length of time, will prevent the action of wood-destroying fungi, insect borers, and similar destructive agents when the wood has been properly coated or impregnated with it. Normally an arsenic derivative. Chromated Copper Arsenate (CCA) is an example.

Pressure Relief Valve (PRV) — A device mounted on a hot water heater or boiler which is designed to release any high steam pressure in the tank to prevent tank explosions

Pressure-treated wood — Lumber that has been saturated with a preservative

Primer — The first, base coat of paint when a paint job consists of two or more coats. A first coating formulated to seal raw surfaces and holding succeeding finish coats.

Principal — The original amount of the loan, the capital

Primary Mortgage Market — Includes banks, savings and loans, credit unions, and mortgage banks that make mortgage loans directly to borrowers. These lenders sometimes sell their mortgages to lenders such as FNMA in the secondary mortgage market.

Prime Rate — Lowest commercial interest rate charged by a bank on short-term loans to its most credit-worthy customers. Often used as an index for home equity lines of credit

Principal — The amount of debt, not counting interest, left on a loan

Profit and Loss Statement — Financial statement showing sales, expenses, and profits over a period of time. Often a requirement for self-employed borrowers

Property Tax — A government tax based on the market value of a property

Property survey — A survey to determine the boundaries of your property. The cost depends on the complexity of the survey.

P trap — Curved, "U" section of drain pipe that holds a water seal to prevent sewer gasses from entering the home through a fixtures water drain

PUD (Planned Unit Development) — A project or subdivision that includes common property that is owned and maintained by a homeowners' association for the benefit and use of the individual PUD unit owner

Pump mix — Special concrete that will be used in a concrete pump. The mix normally has smaller rock aggregate than regular mix.

Punch list — A list of discrepancies that need to be corrected by the contractor

Punch out — To inspect and make a discrepancy list

Purchase Agreement — Contract signed by buyer and seller stating the terms and conditions under which a property will be purchased

Putty — A type of dough used in sealing glass in the sash, filling small holes and crevices in wood, and for similar purposes

PVC or CPVC — Poly Vinyl Chloride. A type of white or light gray plastic pipe sometimes used for water supply lines and waste pipe

Quarry tile — A man-made or machine-made clay tile used to finish a floor or wall. Generally 6 in. X 6 in. X ¼ in. thick

Quarter round — A small trim molding that has the cross section of a quarter circle

Rabbet — A rectangular longitudinal groove cut in the corner edge of a board or plank.

Radiant heating — A method of heating, which can consist of a forced hot water system with pipes placed in the floor, wall, or ceiling. Also electrically heated panels.

Radiation — Energy transmitted from a heat source to the air around it. Radiators actually depend more on convection than radiation.

Radon — A naturally-occurring, heavier than air, radioactive gas common in many parts of the country. Radon gas exposure is associated with lung cancer. Mitigation measures may involve crawl space and basement venting and various forms of vapor barriers.

Radon system — A ventilation system beneath the floor of a basement and/or structural wood floor and designed to fan exhaust radon gas to the outside of the home

Rafter — Lumber used to support the roof sheeting and roof loads. Made from 2 X 10s and 2 X 12s. The rafters of a flat roof are sometimes called roof joists.

Rafter, hip — A rafter that forms the intersection of an external roof angle.

Rafter, valley — A rafter that forms the intersection of an internal roof angle. The valley rafter is normally made of double 2-inch-thick members.

Rail — Cross members of panel doors or of a sash. Also, a wall or open balustrade placed at the edge of a staircase, walkway bridge, or elevated surface to prevent people from falling off. Any relatively lightweight horizontal element, especially those found in fences (split rail).

Rake — The angled edge of a roof, at the end of a roof where it passes the gable

Rake fascia — The vertical face of the sloping end of a roof eave

Rate Lock (or Lock In) — A lender's guarantee of an interest rate and related points for a set period of time, most often between loan application and loan closing. Protects borrower against a rate increase during that time.

Re-bar — Metal rods used to improve the strength of concrete structures

RESPA (Real Estate Settlement Procedures Act) — Law requiring lenders to give borrowers advance notice of closing costs

Real Property — Land and everything that is permanently affixed to it

Realtor — Real estate professional who is a member of the National Association of Realtors

Re-Amortize — The function to provide a new graduated payment amount as it relates to a new loan amount or a new interest rate

Rescission — The cancellation of a contract. With respect to mortgage refinancing, the law that gives the homeowner three days to cancel a contract in some cases once it is signed if the transaction uses equity in the home as security

Reclamation — The right of the person with title to a property to recover it from the debtor in the event of a bankruptcy

Re-conveyance — The transfer of property back to the owner when a mortgage is fully repaid

Recording Fee — Money paid to an agent for entering the sale of a property into the public records

Refinancing — The process of paying off one loan with the proceeds from a new loan secured by the same property

Repossession (or Foreclosure) — Legal process by which the lender forces the sale of a property because the borrower has not met the mortgage terms

Reflective insulation — sheet material with one or both surfaces emitting low heat, such as aluminum foil. The surfaces reduce heat radiation across air space.

Register — Metal plate where air supply is released or returns; can be used to direct the flow of air

Relative humidity — Amount of water vapor in the atmosphere, expressed as a percentage

Resorcinol — An adhesive high in both wet and dry strength. Used to glue boards or other joints that undergo severe conditions

Retainage — A percent of payment that is held back to ensure a job is completed

Return — Ductwork that leads back to the HVAC unit

Reverse board and batten — Narrow battens nailed vertically to wall framing. Wider boards are nailed over these to form siding.

Ribbon or girt — A 1X4 board horizontally supporting ceiling or second-floor joists. Set within the studs

Ripping — Cutting lumber parallel to the grain

Rolled roofing — A roofing material composed of fiber. It is saturated with asphalt and comes in 36-in.-wide rolls

Roof sheathing — The boards or sheets fastened to the rafters, on which shingles or other roofing material is laid

Rottenstone — An abrasive stone used for rubbing a transparent finish to give it a smooth surface

Rout — The removal of material by cutting or gouging a groove

Rubble masonry — Uncut stones used for foundations and other rough work

Run — In stairs, the front to back width of a step or a flight of stairs

R Value — A measure of insulation. A measure of a materials resistance to the passage of heat. The higher the R value, the more insulating "power" it has. For example, typical new home's walls are usually insulated with 4" of batt insulation with an R value of R-13, and a ceiling insulation of R-30.

Saddle — A small second roof built behind the back side of a fireplace chimney or other part of the roof to divert water around the chimney

Sack mix — The amount of Portland cement in a cubic yard of concrete mix. Five or six sack is generally required in a foundation wall.

Sales contract — A contract between a buyer and seller which should explain: (1) What the purchase includes, (2) What guarantees there are, (3) When the buyer can move in, (4) What the closing costs are, and (5) What recourse the parties have if the contract is not fulfilled or if the buyer cannot get a mortgage commitment at the agreed upon time

Sand float finish — Lime that is mixed with sand, resulting in a textured finish on a wall

Sanitary sewer — A sewer system designed for the collection of waste water from the bathroom, kitchen and laundry drains, and is not necessarily designed to handle storm water

Sash — A single light frame containing one or more lights of glass. The frame that holds the glass in a window, often the movable part of the window

Sash balance — A device, usually operated by a spring and designed to hold a single hung window vent up and in place

Saturated felt — A felt which is impregnated with tar or asphalt

Scab — Short length of board nailed over the joint of two boards that butt end to end

Scantling — Lumber with a cross section from 2X4 in. to 4X4 in.

Scarfing — A joint between two pieces of wood that allows them to be spliced lengthwise.

Schedule (window, door, mirror) — A table on the blueprints that list the sizes, quantities and locations of the windows, doors and mirrors.

Scotia — Hollow molding used for part of a cornice

Scrap out — The removal of all drywall material and debris after the home is installed with drywall.

Scratch coat — The first coat of plaster, which is scratched to form a bond for a second coat

Screed, concrete — To level off concrete to the correct elevation during a concrete pour

Screed, plaster — A small strip of wood, the thickness of the plaster coat, used as a guide for plastering

Scribing — Cutting and fitting woodwork to an irregular surface

Scupper — (1) An opening for drainage in a wall, curb or parapet. (2) The drain in a downspout or flat roof, connected to the downspout.

Sealer — A finishing material, either clear or pigmented, that is applied directly over raw wood for the purpose of sealing the wood surface.

Seasoning — Drying and removing moisture from green wood in order to improve its usability.

Second mortgage — A subordinate mortgage made in addition to a first mortgage.

Secondary mortgage market — The market into which primary mortgage lenders sell the mortgages to obtain funds to originate more new loans. Includes investors like Fannie Mae and Freddie Mac.

Self-sealing shingles — Shingles containing factory-applied strips or spots of self-sealing adhesive

Semigloss paint or enamel — A paint or enamel made so that its coating, when dry, has some luster but is not very glossy. Bathrooms and kitchens are normally painted semi-gloss

Septic system — An on-site waste water treatment system. It has a septic tank which promotes the biological digestion of the waste and a drain field which is designed to let the left over liquid soak into the ground. Septic systems and permits are sized by the number of bedrooms in a house.

Service entrance panel — Main power cabinet where electricity enters a home wiring system

Service equipment — Main control gear at the service entrance, such as circuit breakers, switches, and fuses

Service lateral — Underground power supply line

Servicing (or Loan Administration) — The collection of mortgage payments from borrowers and related responsibilities, such as handling escrows for property tax and insurance, foreclosing on defaulted loans, and remitting payments to investors

Settlement (or Closing) — Meeting between the buyer, seller, and closing agent at which property and funds legally change hands

Settlement Sheet — The computation of costs payable at closing which determines the seller's net proceeds and the buyer's net payment

Setback Thermostat — A thermostat with a clock which can be programmed to come on or go off at various temperatures and at different times of the day/week. Used as the heating or cooling system thermostat

Settlement — Shifts in a structure, caused by freeze-thaw cycles underground

Sewage ejector — A pump used to 'lift' waste water to a gravity sanitary sewer line. Used in basements and other locations which are situated below the level of the side sewer

Sewer lateral — The portion of the sanitary sewer which connects the interior waste water lines to the main sewer lines. The side sewer is buried in several feet of soil and runs from the house to the sewer line. It is 'owned' by the sewer utility, must be maintained by the owner and may only be serviced by utility approved contractors. Sometimes called side sewer.

Sewer stub — The junction at the municipal sewer system where the home's sewer line is connected

Sewer tap — The physical connection point where the home's sewer line connects to the main municipal sewer line

Shake — A wood roofing material, normally cedar or redwood. Produced by splitting a block of the wood along the grain line. Modern shakes are sometimes machine sawn on one side. See shingle.

Shear block — Plywood that is face nailed to short (2 X 4's or 2 X 6's) wall studs (above a door or window, for example). This is done to prevent the wall from sliding and collapsing.

Sheathing, sheeting — The structural wood panel covering, normally OSB or plywood, used over studs, floor joists or rafters/trusses of a structure

Shed roof — A roof containing only one sloping plane

Sheet metal work — All components of a house employing sheet metal, such as flashing, gutters, and downspouts

Sheet metal duct work — The heating system. Round or rectangular metal pipes and sheet metal (for Return Air) and installed for distributing warm (or cold) air from the furnace to rooms in the home

Sheet rock, drywall, wall board, or gypsum — A manufactured panel made out of gypsum plaster and encased in a thin cardboard. Usually ½ in. thick and 4 ft. X 8 ft. or 4 ft. x 12 ft. in size. The 'joint compound' 'green board' type drywall has a greater resistance to moisture than regular (white) plasterboard and is used in bathrooms and other "wet areas".

Shim — A small piece of scrap lumber or shingle, often wedge shaped, which when forced behind a furring strip or framing member forces it into position. Also used when installing doors and placed between the door jamb legs and 2 X 4 door trimmers. Metal shims are wafer 1½ in. X 2 in. sheet metal of various thicknesses used to fill gaps in wood framing members, especially at bearing point locations.

Shingles — Roof covering of asphalt. Asbestos, wood, tile, slate, or other material cut to stock lengths, widths, and thicknesses

Shingles, siding — Various kinds of shingles, used over sheathing for exterior wall covering of a structure

Short circuit — A situation that occurs when hot and neutral wires come in contact with each other. Fuses and circuit breakers protect against fire that could result from a short.

Shutter — Lightweight louvered decorative frames in the form of doors located on the sides of a window. Some shutters are made to close over the window for protection.

Side sewer — The portion of the sanitary sewer which connects the interior waste water lines to the main sewer lines. The side sewer is buried in several feet of soil and runs from the house to the sewer line. It is 'owned' by the sewer utility, must be maintained by the owner and may only be serviced by utility approved contractors. Sometimes called sewer lateral.

Siding — The finished exterior covering of the outside walls of a frame building

Siding, (lap siding) — Slightly wedge-shaped boards used as horizontal siding in a lapped pattern over the exterior sheathing. Varies in butt thickness from ½ to ¾ in. and in widths up to 12 in.

Sill — (1) The 2 X 4 or 2 X 6 wood plate framing member that lays flat against and bolted to the foundation wall (with anchor bolts) and upon which the floor joists are installed. Normally the sill plate is treated lumber. (2) The member forming the lower side of an opening, as a door sill or window sill

Sill cock — An exterior water faucet (hose bib)

Sill plate (mudsill) — Bottom horizontal member of an exterior wall frame which rests on top a foundation, sometimes called mudsill. Also sole plate, bottom member of an interior wall frame

Sill seal — Fiberglass or foam insulation installed between the foundation wall and sill (wood) plate. Designed to seal any cracks or gaps

Simple Interest — Interest computed only on the principal balance

Single hung window — A window with one vertically sliding sash or window vent

Skylight — A more or less horizontal window located on the roof of a building

Slab, concrete — Concrete pavement, i.e. driveways, garages, and basement floors

Slab, door — A rectangular door without hinges or frame

Slab on grade — A type of foundation with a concrete floor which is placed directly on the soil. The edge of the slab is thicker and acts as the footing for the walls.

Slag — Concrete cement that sometimes covers the vertical face of the foundation material

Sleeper — A wood member embedded in concrete, as in a floor, that serves to support and to fasten the subfloor or flooring

Sleeve(s) — Pipe installed under the concrete driveway or sidewalk, and that will be used later to run sprinkler pipe or low voltage wire

Slope — The incline angle of a roof surface, given as a ratio of the rise (in inches) to the run (in feet). See also pitch.

Slump — The "wetness" of concrete. A 3 in. slump is dryer and stiffer than a 5 in. slump.

Soffit — The area below the eaves and overhangs. The underside where the roof overhangs the walls. The underside of an overhanging cornice

Soil pipe — A large pipe that carries liquid and solid wastes to a sewer or septic tank

Soil stack — A plumbing vent pipe that penetrates the roof.

Sole plate — The bottom, horizontal framing member of a wall that is attached to the floor sheeting and vertical wall studs

Solid bridging — A solid member placed between adjacent floor joists near the center of the span to prevent joists or rafters from twisting

Sonotube — Round, large cardboard tubes designed to hold wet concrete in place until it hardens

Sound attenuation — Sound proofing a wall or subfloor, often through the use of fiberglass insulation

Space heat — Heat supplied to the living space, for example, to a room or the living area of a building

Spacing — The distance between individual members or shingles in building construction

Span — The clear distance that a framing member carries a load without support between structural supports. The horizontal distance from eaves to eaves

Spec home — A house built before it is sold. The builder speculates that he can sell it at a profit.

Specifications or Specs — A narrative list of materials, methods, model numbers, colors, allowances, and other details which supplement the information contained in the blue prints. Written elaboration in specific detail about construction materials and methods. Written to supplement working drawings

Splash block — Portable concrete (or vinyl) channel placed beneath an exterior sill cock (water faucet) or downspout in order to receive roof drainage from downspouts and to divert it away from the building

Square — A unit of measure-100 sq. ft. — applied to roofing and siding material. Also, a situation that exists when two elements are at right angles to each other. Also a tool for checking this

Square-tab shingles — Shingles on which tabs are all the same size and exposure

Squeegie — Fine pea gravel used to grade a floor (normally before concrete is placed)

Stack (trusses) —To position trusses on the walls in their correct location

Standard practices of the trade(s) — One of the more common basic and minimum construction standards. This is another way of saying that the work should be done in the way it is normally done by the average professional in the field.

Starter strip — Asphalt roofing applied at the eaves that provides protection by filling in the spaces under the cutouts and joints of the first course of shingles

Stair carriage or stringer — Supporting member for stair treads. A 2 X 12 in. plank notched to receive the treads; sometimes called a "rough horse"

Stair landing — A platform between flights of stairs or at the termination of a flight of stairs. Often used when stairs change direction. Normally no less than 3 ft. X 3 ft. square

Stair rise — The vertical distance from stair tread to stair tread (and not to exceed 7 ½ in.)

Static vent — A vent that does not include a fan

STC (Sound Transmission Class) — The measure of sound stopping of ordinary noise

Steel inspection — A municipal and/ or engineers inspection of the concrete foundation wall, conducted before concrete is poured into the foundation panels. Done to insure that the rebar (reinforcing bar), rebar nets, void material, beam pocket plates, and basement window bucks are installed and wrapped with rebar and complies with the foundation plan

Step flashing — Flashing application method used where a vertical surface meets a sloping roof plane. 6" X 6" galvanized metal bent at a 90 degree angle, and installed beneath siding and over the top of shingles. Each piece overlaps the one beneath it the entire length of the sloping roof (step by step).

Stick built — A house built without prefabricated parts. Also called conventional building

Stile — An upright framing member in a panel door.

Stool — The flat molding fitted over the window sill between jambs and contacting the bottom rail of the lower sash. Also another name for toilet

Stop box — Normally a cast iron pipe with a lid (around 5 in. diameter) that is placed vertically into the ground, situated near the water tap in the yard, and where a water cut-off valve to the home is located (underground). A long pole with a special end is inserted into the curb stop to turn off/on the water.

Stop Order — A formal, written notification to a contractor to discontinue some or all work on a project for reasons such as safety violations, defective materials or workmanship, or cancellation of the contract

Stops — Moldings along the inner edges of a door or window frame. Also valves used to shut off water to a fixture.

Stop valve — A device installed in a water supply line, near a fixture, that permits an individual to shut off the water supply to one fixture without interrupting service to the rest of the system

Storm sash or storm window — An extra window placed outside of an existing one, as additional protection against cold weather

Storm sewer — A sewer system designed to collect storm water and is separated from the waste water system

Story — That part of a building between any floors or between the floor and roof

Strike — The plate on a door frame that engages a latch or dead bolt

String, stringer — A timber or other support for cross members in floors or ceilings. In stairs, the supporting member for stair treads. A 2 X 12 inch plank notched to receive the treads

Strip flooring — Wood flooring consisting of narrow, matched strips

Structural floor — A framed lumber floor that is installed as a basement floor instead of concrete. This is done on very expansive soils.

Stub, stubbed — To push through

Stucco — Refers to an outside plaster finish made with Portland cement as its base

Stud — A vertical wood framing member, also referred to as a wall stud, attached to the horizontal sole plate below and the top plate above. Normally 2 X 4's or 2 X 6's, 8' long (sometimes 92 in.). One of a series of wood or metal vertical structural members placed as supporting elements in walls and partitions

Stud framing — A building method that distributes structural loads to each of a series of relatively lightweight studs. Contrasts with post-and-beam

Stud shoe — A metal, structural bracket that reinforces a vertical stud. Used on an outside bearing wall where holes are drilled to accommodate a plumbing waste line.

Subfloor — The framing components of a floor to include the sill plate, floor joists, and deck sheeting over which a finish floor is to be laid

Sump — Pit or large plastic bucket/ barrel inside the home designed to collect ground water from a perimeter drain system

Sump pump —A submersible pump in a sump pit that pumps any excess ground water to the outside of the home

Suspended ceiling — A ceiling system supported by hanging it from the overhead structural framing

Sway brace — Metal straps or wood blocks installed diagonally on the inside of a wall from bottom to top plate, to prevent the wall from twisting, racking, or falling over "domino" fashion

Switch — A device that completes or disconnects an electrical circuit

Survey — A measurement of land, prepared by a licensed surveyor, showing a property's boundaries, elevations, improvements, and relationship to surrounding tracts

T & G, tongue and groove — A joint made by a tongue (a rib on one edge of a board) that fits into a corresponding groove in the edge of another board to make a tight flush joint. Typically, the subfloor plywood is T & G.

Tab — The exposed portion of strip shingles defined by cutouts

Tail beam — A relatively short beam or joist supported in a wall on one end and by a header at the other

Take off — The material necessary to complete a job

Taping — The process of covering drywall joints with paper tape and joint compound

Tax lien — Claim against a property for unpaid taxes

Tax sale — Public sale of property by a government authority as a result of non-payment of taxes

T bar — Ribbed, "T" shaped bars with a flat metal plate at the bottom that are driven into the earth. Normally used chain link fence poles, and to mark locations of a water meter pit

Teco — Metal straps that are nailed and secure the roof rafters and trusses to the top horizontal wall plate. Sometimes called a hurricane clip

Tee — A "T" shaped plumbing fitting

Tempered — Strengthened. Tempered glass will not shatter nor create shards, but will "pelletize" like an automobile window. It is required in tub and shower enclosures and locations, entry door glass and sidelight glass, and in a windows when the window sill is less than 16 in. to the floor.

Tenon — Projection at the end of the board that inserts into a mortise

Termites — Wood eating insects that superficially resemble ants in size and general appearance, and live in colonies

Termite shield — A shield of galvanized metal, placed in or on a foundation wall or around pipes to prevent the passage of termites

Terra cotta — A ceramic material molded into masonry units

Thermoply ™ — Exterior laminated sheathing nailed to the exterior side of the exterior walls. Normally ¼ in. thick, 4 X 8 or 4 x 10 sheets with an aluminumized surface

Three-dimensional shingles — also called "architectural shingles". Laminated shingles. Shingles that have added dimensionality because of extra layers or tabs, giving a shake-like appearance

Threshold — The bottom metal or wood plate of an exterior door frame. They are adjustable to keep a tight fit with the door slab.

Tie beam, or collar beam — Ties together the principal rafters of a roof

Tieback member — A timber perpendicular to a retaining wall that ties it to a deadman buried in the ground

Time and materials contract — A construction contract which specifies a price for different elements of the work such as cost per hour of labor, overhead, profit. A contract which may not have a maximum price, or may state a 'price not to exceed'.

Tinner — Another name for the heating contractor.

Tip up — The downspout extension that directs water (from the home's gutter system) away from the home. They typically swing up when mowing the lawn.

Title — Evidence (in the form of a certificate or deed) of a person's legal right to ownership of a property.

Title Company — A company that insures title to property

Title Insurance — Insurance which protects the lender (lender's policy) or the buyer (owner's policy) against loss due to disputes over ownership of a property

Title Search — Examination of municipal records to ensure that the seller is the legal owner of a property and that no liens or other claims exist against the property

Toenailing — To drive a nail in at a slant. Method used to secure floor joists to the plate

Top chord — The upper or top member of a truss.

Top plate —Top horizontal member of a frame wall supporting ceiling joists, rafters, or other members

Transfer Tax — Tax paid when title passes from one owner to another. Not applicable in all jurisdictions

Transmitter (garage door) — The small, push button device that causes the garage door to open or close

Trap — A plumbing fitting that holds water to prevent air, gas, and vermin from backing up into a fixture

Tray ceiling — Raised area in a ceiling that looks like a small vaulted ceiling

Tread —The walking surface board in a stairway on which the foot is placed

Treated lumber — A wood product which has been impregnated with chemical pesticides such as CCA (Chromated Copper Arsenate) to reduce damage from wood rot or insects. Often used for the portions of a structure which are likely to be in contact with soil and water. Wood may also be treated with a fire retardant.

Trim (plumbing, heating, electrical) — The work that the "mechanical" contractors perform to finish their respective aspects of work, and when the home is nearing completion and occupancy

Trim, Interior — The finish materials in a building, such as moldings applied around openings (window trim, door trim) or at the floor and ceiling of rooms (baseboard, cornice, and other moldings). Also, the physical work of installing interior doors and interior woodwork, to include all handrails, guardrails, stair way balustrades, mantles, light boxes, base, door casings, cabinets, countertops, shelves, window sills and aprons

Trim, Exterior — The finish materials on the exterior a building, such as moldings applied around openings (window trim, door trim), siding, windows, exterior doors, attic vents, crawl space vents, shutters. Also, the physical work of installing these materials

Trimmer — The vertical stud that supports a header at a door, window, or other opening

Troweling — process of smoothing concrete, used after floating

Trust Account — Account maintained by a broker or escrow company to handle all money collected for clients

Trustee — Someone given legal responsibility to hold property in the best interest of another

Truss — An engineered and manufactured roof support member with "zig-zag" framing members. Does the same job as a rafter but is designed to have a longer span than a rafter

Truth-In-Lending Act — Federal law requiring written disclosure of the terms of a mortgage by a lender to a prospective borrower within three business days of application

Tub trap — Curved, "U" shaped section of a bath tub drain pipe that holds a water seal to prevent sewer gasses from entering the home through tubs water drain

Turnkey — A term used when the subcontractor provides all materials (and labor) for a job

Turpentine — A petroleum, volatile oil used as a thinner in paints and as a solvent in varnishes

UL (Underwriters' Laboratories) — An independent testing agency that checks electrical devices and other components for possible safety hazards

Undercoat — A coating applied prior to the finishing or top coats of a paint job. It may be the first of two or the second of three coats. Sometimes called the Prime coat

Underground plumbing — The plumbing drain and waste lines that are installed beneath a basement floor

Underlayment — A ¼ in. material placed over the subfloor plywood sheeting and under finish coverings, such as vinyl flooring, to provide a smooth, even surface. Also a secondary roofing layer that is waterproof or water-resistant, installed on the roof deck and beneath shingles or other roof-finishing layer

Underwriting — The process of verifying data and evaluating a loan for approval

Union — A plumbing fitting that joins pipes end-to-end so they can be dismantled

Usury — Interest charged in excess of the legal rate established by law

Utility easement — The area of the earth that has electric, gas, or telephone lines. These areas may be owned by the homeowner, but the utility company has the legal right to enter the area as necessary to repair or service the lines.

Valley — The "V" shaped area of a roof where two sloping roofs meet. Water drains off the roof at the valleys.

Valley flashing — Sheet metal that lays in the "V" area of a roof valley

Valuation — An inspection carried out for the benefit of the mortgage lender to ascertain if a property is a good security for a loan

Valuation fee — The fee paid by the prospective borrower for the lender's inspection of the property. Normally paid upon loan application

Vapor barrier — A building product installed on exterior walls and ceilings under the drywall and on the warm side of the insulation. It is used to retard the movement of water vapor into walls and prevent condensation within them. Normally, polyethylene plastic sheeting is used.

Variable rate — An interest rate that will vary over the term of the loan

Veneer — Extremely thin sheets of wood. Also a thin slice of wood or brick or stone covering a framed wall

Vent — A pipe or duct which allows the flow of air and gasses to the outside. Also, another word for the moving glass part of a window sash, i.e. window vent

Vermiculite — A mineral used as bulk insulation and also as aggregate in insulating and acoustical plaster and in insulating concrete floors

Veterans Administration (VA) — A federal agency that insures mortgage loans with very liberal down payment requirements for honorably discharged veterans and their surviving spouses

Visqueen — A 4 mil or 6 mil plastic sheeting

Void — Cardboard rectangular boxes that are installed between the earth (between caissons) and the concrete foundation wall. Used when expansive soils are present

Voltage — A measure of electrical potential. Most homes are wired with 110 and 220 volt lines. The 110 volt power is used for lighting and most of the other circuits. The 220-volt power is usually used for the kitchen range, hot water heater and dryer.

Wafer board — A manufactured wood panel made out of 1"- 2" wood chips and glue. Often used as a substitute for plywood in the exterior wall and roof sheathing

Waiver — Voluntary relinquishment or surrender of some right or privilege

Walk-Through — A final inspection of a new home to check for problems that may need to be corrected before closing

Wall out — When a painter pray paints the interior of a home

Warping — Any distortion in a material

Warranty — In construction there are two general types of warranties. One is provided by the manufacturer of a product such as roofing material or an appliance. The second is a warranty for the labor. For example, a roofing contract may include a 20 year material warranty and a 5 year labor warranty. Many new homebuilders provide a one year warranty. Any major issue found during the first year should be communicated to the builder immediately.

Waste pipe and vent — Plumbing plastic pipe that carries waste water to the municipal sewage system

Water board — Water resistant drywall to be used in tub and shower locations. Normally green or blue colored

Water closet — Another name for toilet

Water meter pit (or vault) — The box /cast iron bonnet and concrete rings that contains the water meter

Water-repellent preservative — A liquid applied to wood to give the wood water repellant properties

Water table — The location of the underground water, and the vertical distance from the surface of the earth to this underground water

Water tap — The connection point where the home water line connects to the main municipal water system

W C — An abbreviation for water closet (toilet)

Weatherization — Work on a building exterior in order to reduce energy consumption for heating or cooling. Work involving adding insulation, installing storm windows and doors, caulking cracks and putting on weather-stripping

Weatherstrip — Narrow sections of thin metal or other material installed to prevent the infiltration of air and moisture around windows and doors

Weep holes — Small holes in storm window frames that allow moisture to escape

Whole house fan — A fan designed to move air through and out of a home and normally installed in the ceiling

Wind bracing — Metal straps or wood blocks installed diagonally on the inside of a wall from bottom to top plate, to prevent the wall from twisting, racking, or falling over "domino" fashion

Window buck — Square or rectangular box that is installed within a concrete foundation or block wall. A window will eventually be installed in this "buck" during the siding stage of construction

Window frame — The stationary part of a window unit; window sash fits into the window frame

Window sash — The operating or movable part of a window; the sash is made of window panes and their border

Wire nut — A plastic device used to connect bare wires together

Wonderboard ™ — A panel made out of concrete and fiberglass used as a ceramic tile backing material. Commonly used on bathtub decks

Worker's Compensation — This is a policy or endorsement covering the contractor, subcontractor, and others who will be working on the subject property. This policy is typically provided by the contractor, thus the contractor should be named as the insured. It states that in cases in which worker's comp insurance is not required or the borrower is acting as his or her own general contractor, a waiver is required to be executed.

Wrapped drywall — Areas that get complete drywall covering, as in the doorway openings of bifold and bypass closet doors

Wrap-Around Mortgage — Loan arrangement in which an existing loan is combined with a new loan, resulting in an interest rate somewhere between the old rate and the current market rate

Y — A "Y" shaped plumbing fitting

Yard of concrete — Concrete is 3 ft. X 3 ft. X 3 ft. in volume, or 27 cubic ft. One cubic yd. of concrete will pour 80 square feet of 3 ½ in. sidewalk or basement/garage floor.

Yoke — The location where a home's water meter is sometimes installed between two copper pipes, and located in the water meter pit in the yard

Z-bar flashing — Bent, galvanized metal flashing that is installed above a horizontal trim board of an exterior window, door, or brick run. It prevents water from getting behind the trim/brick and into the home.

Zone — The section of a building that is served by one heating or cooling loop because it has noticeably distinct heating or cooling needs. Also, the section of property that will be watered from a lawn sprinkler system

Zone valve — A device placed near the heater or cooler, which controls the flow of water or steam to parts of the building; it is controlled by a zone thermostat

Zoning Ordinances (or Zoning Regulations) — Local law establishing building codes and usage regulations for properties in a specified area

Index

A

B

C

O

P

Q

R

S

U

W